PRESERVING THE CATCH

PRESERVING THE CATCH

Erling Stuart

Stackpole Books

Published by
STACKPOLE BOOKS
Cameron and Kelker Streets
P.O. Box 1831
Harrisburg, PA 17105

Published simultaneously in Scarborough, Ontario, Canada
by Thomas Nelson & Sons, Ltd.

Printed in the U.S.A.

Library of Congress Cataloging in Publication Data

Stuart, Erling.
Preserving the catch.

Includes index.
1. Fishery products—Preservation. 2. Fishes. Dressing of. I. Title.
TX612.F5S83 641.4'94 81-21495
ISBN 0-8117-1285-0 AACR2

Contents

Acknowledgments

Many people and organizations have helped in the production of this book. The people in the fishing industry, both in Alaska and in Washington, and the people I have met on different fishing boats are too numerous to mention individually, but they are all due a sincere thank you. The scientific publications of the U.S. Department of the Interior and the National Marine Fisheries have been invaluable, particularly for chapter 4. To all of the authors of the many books I have read over the years, I am grateful.

I am indebted to Professor William Matchett of the University of Washington for his valuable criticism of the first draft of the manuscript. Any errors that may have crept in to subsequent revisions are mine, not his. To my grandfather, I owe many thanks for showing me "how they did it in Norway." And, finally, I thank my wife, Tracy Mork, for her patience and encouragement in helping me finish this project. Not only did she help with the seemingly endless typing, but her drawings made clear what I may have muddled through words.

Introduction

Ever since man began harvesting fish from the sea, he has developed different techniques for preserving his catch. The climate, the type of fish, and the way the fish were to be used all had a part in determining how the fish were to be preserved.

Some people caught fresh fish every day and relied on very basic methods to keep fish for only a few hours. Damp leaves were often used to keep the fish cool until they could be eaten. Other societies relied upon more migratory fish and developed methods to keep them from one season to the next. Salting and drying are simple methods that were popular for centuries.

Recent research in Controlled Atmosphere storage may prove useful in home storage someday. In this system, nitrogen, carbon dioxide, or inert gases are injected into the container holding the fish to displace the oxygen. Bacteria and other organisms suffocate, and the effects of oxygen on fish are eliminated.

Radiation is being studied to determine if it can provide a low cost means of storing fish. This type of preservation keeps fish by killing harmful bacteria and other organisms with microwave radiation.

Today, fish are preserved in different ways for many reasons. Some people preserve fish to maintain a link with the past. Tradition and the memories of the family working together to put up the food supply provide a strong motivation to keep food by some means other than simply refrigerating or freezing it. Fish can be smoked for the delicious flavor as much as for its keeping qualities. Some people choose to can food rather than depend on more mechanical (and energy-intensive) methods such as freezing. Often, canned fish proves to be more versatile in the ways it can be prepared than fish that is merely frozen. Certainly canned fish is more portable. No one would want to pickle all their fish, but for variety, pickled fish is hard to surpass.

The guiding principle behind all methods of preserving fish is the creation of an environment that is unfavorable to the growth and development of harmful organisms. In refrigeration and freezing, cold makes conditions inhospitable to bacteria. In canning, heat is responsible for killing the microbes. Dehydration and salting are similar in that they retard the destruction caused by bacteria, molds, and yeasts. And pickling creates an environment that is too acidic for these organisms.

Biological reactions are not the only consideration in storing fish over a period of time. Chemical reactions that damage fish can also be controlled by varying the ambient conditions. Fish stored at high temperatures (above 65°) are vulnerable to proteolysis, rancidity, autolysis, and oxidation. Bright light and the presence of atmospheric oxygen and metal ions also increase the rate of oxidation.

Whatever method you choose to keep your fish, follow the guidelines outlined in this book. Some methods invite experimentation. The rules are not hard and fast. Other methods, particularly canning, require strict adherence to definite procedures. Use common sense. Fish that are allowed to become warm are not going to be in prime condition. It should go without saying, keep everything as clean as possible in any preserving method.

The ability of any preserving system to kill bacteria is directly related to the number of bacteria initially present. The situation is similar to trying to put out a big fire with a small extinguisher. Never assume that any germs present on the fish will be killed later in the preserving process. It simply won't happen. The average fish containing an average number of bacteria can be successfully preserved. A partially decomposed or contaminated fish cannot.

The most important reason for preserving fish at home is that *you* control the quality of the final product. You are responsible for the fish's wholesomeness, flavor, and nutrition. The way your fish looks—whether it is simply food to nourish the body, or whether it nourishes the senses as well—is up to you. You alone will determine the value of your food.

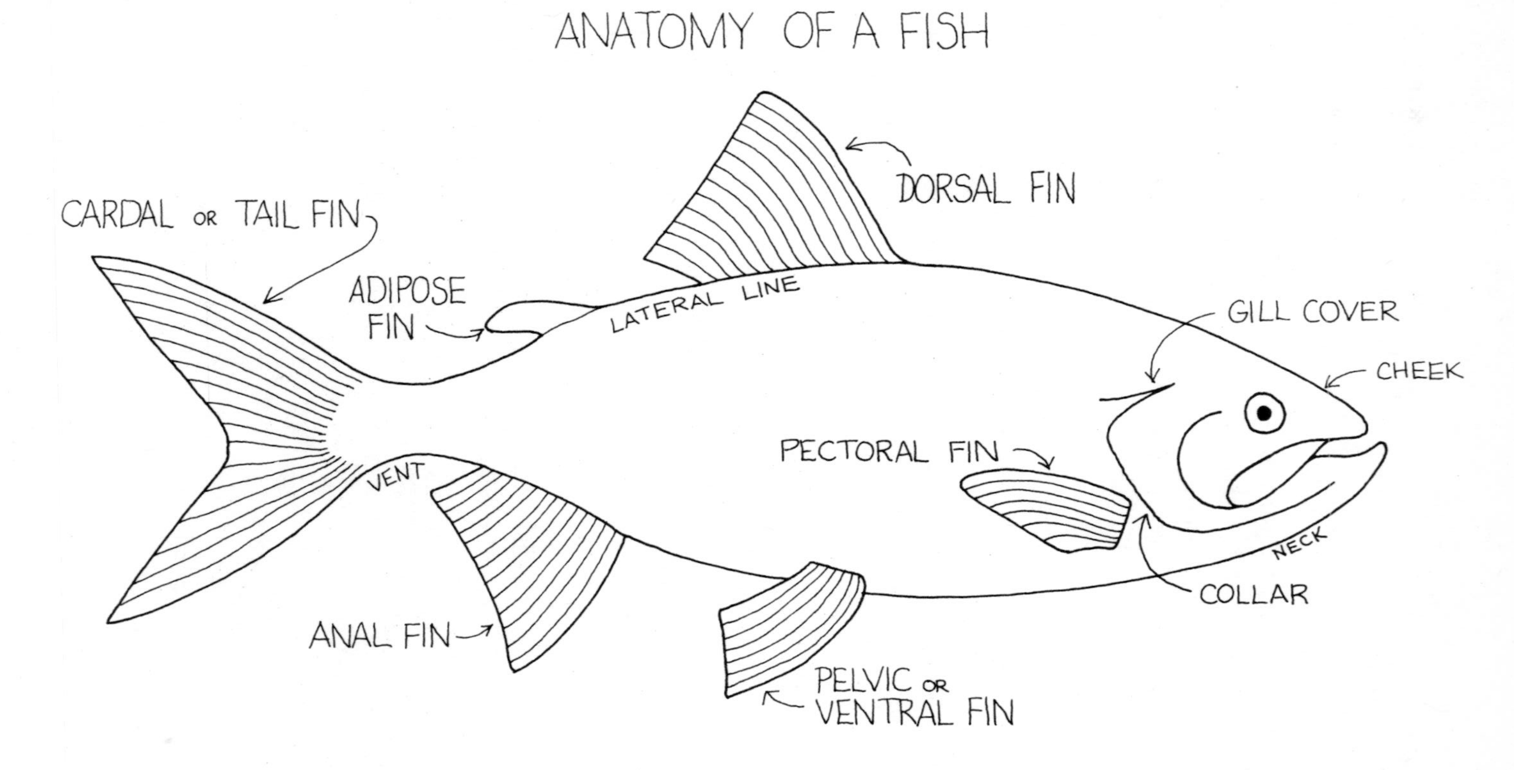

Fig. 1. The anatomy of a fish.

1

Fresh and Frozen Fish

A shiny, bright fish fresh from the water is a thing of beauty, and the goal of the fisherman, whether fishing for sport or for subsistence, is to preserve the fish while it is still fresh and in its best condition. Each fish should be handled individually and killed quickly so that it does not become bruised from thrashing about in the boat. Ideally, it should then be taken directly to the kitchen and prepared for the next meal. However, it is not always possible or convenient to eat the fish as soon as it is caught, so some method of storing it must be adopted.

Another reason to keep the fish from flopping around in the boat is to allow rigor mortis to set in quickly. Rigor mortis, the temporary stiffening of muscles after death, plays a role in preserving the fish. Decomposition and the breakdown of tissue both occur at a slower rate during rigor mortis. In addition, fish that are chilled right away, as well as fish that are not strenuously worked right before death, go through a longer period of rigor mortis. Butchering the fish immediately also extends the period of rigor mortis.

GUTTING THE FISH

The first step in preserving the freshly caught fish is to remove the internal organs to prevent autolysis and putrefaction. Autolysis is the process in which the digestive enzymes in the stomach of a feeding fish begin to digest the muscle tissue itself. In just a few hours these enzymes "burn" their way through the organs of the fish and reach the body tissues. Gutting the fish also helps to bleed it.

Putrefaction occurs when bacteria release their own digestive enzymes to break down body tissue. The liquefied tissue is absorbed by the bacteria and waste products (toxins) are released. The number of bacteria on a fish or in its organs is a major problem. Certain bacteria double their number every fifteen minutes, or nearly one hundred generations in twenty-four hours.

Both autolysis and putrefaction are accelerated at warm temperatures. Icing the body cavity or salting the fish lightly after the initial cleaning will minimize these processes.

IS THE FISH FRESH?

Not everyone can enjoy the luxury of catching his own fish fresh from the water. But even if you must buy your fish in a market, there are several ways to determine if it is fresh. The eyes should be clear and bulging, and the gills should be deep red. As the fish begins to age, the eyes become cloudy and the cornea opaque. The color of the gills changes from red to pink to gray to brown. Then in the final stage, as the bilirubin (a reddish-yellow pigment in the blood) breaks down, the gills turn greenish. The gills of a fish that has been lying in melted ice for a long period of time, however, will have become bleached and almost colorless.

Another way to determine freshness is to check the color of

the meat. When the head is cut off or the gills removed, the meat will be visible at the neck. If the fish is whole, you can take a small nick out of the fish just above the tail. The meat should be a deep color depending on the kind of fish. If it is pale, it is probably not fresh. Salmon meat varies in color from red to orange-pink depending on the species, while the meat of white fish, such as halibut and cod, should have a slight bluish tint. Shiny white flesh indicates that the fish has been soaking in melted ice and thus has lost some of its nutritional qualities.

Knowing the breeding seasons of fish is also helpful. Fish are usually under stress just before and after spawning, and the effects of this stress will show in the meat itself. In salmon, for example, greenish or black stripes and blotches, called water marks, become visible as the fish begin their journey through fresh water to spawn. Many people would be very surprised if they could see the whole salmon mottled with these water marks, rather than the already cut "prime" steaks they are buying.

The meat of fresh fish that has been properly handled will be firm and have no soft spots that indicate bruises, and it will not appear to be separating from the bones. In a gutted fish the carcass will still be round when it is laid on its side, leaving a cavity where the organs were. The skin will be smooth and moist, the scales firmly attached, and there will be no odor. Only after the fish begins to age will it develop the characteristic fishy smell. If the fish is old, the slime will be thick and ropy.

To stay fresh, the fish must remain chilled from the moment it is taken from the water until the time it is actually prepared. Any appreciable rise in temperature will diminish the quality. Exposure to sunlight is particularly harmful. Light increases the rate of oxidation and causes the color of the meat to fade. While color is mainly a matter of visual appeal, the lack of color is a significant factor in judging quality. Light also breaks down the riboflavin (vitamin B_2) in the muscle.

Fish should always be cleaned as soon as possible after they are caught. Wash the outsides of the fish before gutting them to remove surface bacteria.

CLEANING SALMON AND OTHER ROUND-SHAPED FISH

To clean round-shaped fish such as salmon, cut off the head by starting the cut at the throat (the ventral side of the fish) ahead of the collar, just behind the gills. Follow the thick, cartilaginous collar in a half-circle to the top of the head, leaving a moderately long nape (the top of the head that is smooth and also full of cartilage). Next, facing the belly, make a slit between the pelvic fins from the anal vent to an inch and a half before the throat. The reason for stopping here is that it makes the fish much easier to handle. If it is simply cut in half, you will have two floppy, unwieldy pieces of meat.

Now from the head end, make two small cuts on either side of the gills that remain and remove them. Then remove the organs through the slit in the belly. Save any eggs (roe) from female fish to use as caviar or fish bait (see chapter 8). With the tip of the knife, slit the membrane along the backbone and scrape the blood line (actually the kidney) until the bone is clean.

Rinse the fish and place it in a bucket or pan of salted ice water (one-half cup of *uniodized* salt dissolved in one gallon of water) for five minutes. The cold water helps draw the blood from the veins for a nicer looking piece of meat, and the salt firms up the flesh by drawing some of the fluids from the tissues. Four teaspoons of vinegar can also be added to each gallon of saline solution to dissolve the slime on the fish.

When preparing the saline solution, dissolve all the salt in half of the water, then add it to the remaining half. Otherwise, even though the salt has dissolved completely, the concentration of the solution will be stronger at the bottom of the container than at the top. Water temperature has no impact on the amount of salt that can be dissolved. Just as much salt can be dispersed in a gallon of cold water as in a gallon of hot water.

After removing the fish from the ice water, use a round-end knife to scrape out any blood that remains. The inside of the belly has many small blood vessels running parallel to the ribs. Use the knife to scrape the blood from the ventral side of the fish to the dorsal side. Be careful not to cut or bruise the meat while scraping.

The two small clots of blood on either side of the neck—consisting of about one teaspoon each—should also be removed by pressing with the knife. These clots are similar to sweetbreads, and they sour quickly if the fish is mishandled. At this stage of the cleaning process, there should be no traces of viscera or membranes along the body cavity.

Scrape the slime off the fish from head to tail. If you are going to freeze the fish, do not scrape the scales. They will act as a protective layer and prevent freezer burn. When the fish is thawed, the scales can be removed by scraping from tail to head.

After cleaning the fish, dip it in a tub of cold water to which household chlorine bleach (hypochlorite) has been added. The ratio is one tablespoon chlorine bleach to four gallons of water. The chlorine is effective in killing any bacteria on the fish and preventing the growth of new bacteria. The solution also removes any remaining slime.

To cut steaks from a salmon or other round-shaped fish, begin by severing the backbone. Then turn the fish on its side and complete the cut. (See Fig. 2.)

Skinning the Fish

If you want to skin the fish, now is the time to do it. Fish with large scales, such as carp, are often skinned in order to avoid the messy job of scaling. Catfish and eels are skinned because their skin is very tough. It is also a good idea to skin fish that are to be breaded.

To skin salmon fillets (see chapter 2 for a discussion of filleting salmon), lay the fillet on the table skin side down. Place the tail end of the fish toward you, head straight away. Hold the tail against the table with your left hand, and with your right hand cut the flesh away from the skin for a distance of about one inch. Continue holding the tail with one hand, and while holding the knife at a downward angle, push the meat off the skin. Don't try to slice the meat free.

CLEANING, SKINNING, AND FILLETING FLOUNDER AND OTHER FLAT FISH

In contrast to round-shaped fish, which can be cleaned without skinning, skinning is the first step in cleaning flat fish such as flounder. Wash the fish to remove any bacteria and slime. Also remove the skin. Laying the fish on the table with the tail toward you, head straight away. (See Fig. 3.) Hold the tail in one hand and cut through the skin at the tail. Be careful not to cut into the

Fig. 2. To cut steaks from salmon or other round-shaped fish, begin by severing the backbone. Turn the fish on its side to complete the cut.

Fig. 3. The first step in skinning flounder or other flat fish is to slip the knife under the skin at the tail and slide it forward above the fins.

meat. Lift the skin with your other hand and pull it forward to the head. (See Fig. 4.) Continue to lift as you pull. Turn the fish over and repeat the process.

To keep a firm grip on the slippery tail, wrap it in a cloth towel or wear cotton gloves. Use a pair of pliers to grip the skin when pulling it off.

Flatfish are cleaned somewhat differently than round fish. In flat fish the viscera is near the head rather than along the backbone. To gut a flat fish, simply cut off the head around the collar and pull out any organs that are still attached to the body. If you are going to fillet the fish, you will not have to bother removing the organs.

It is a simple matter to fillet flat fish. Place the fish on the worktable so that the head is at your left and the tail at your right. The backbone should be parallel to the edge of the worktable. Look for the line, from head to tail on the skinned fish, that delineates the backbone. Run the point of the knife along the line to separate the two fillets that are situated on either side of it. (See Fig. 5.) Work the blade of the knife under the fillet, beginning at the vertebrae, and move it toward the edge of the fish—that is, toward you. Now turn the fish so that the head is at your right and repeat the process for the second fillet. Turn the fish over and repeat for the other two fillets.

Fish that have been skinned must be handled carefully. Since fish have little connective tissue, the skin plays an important role

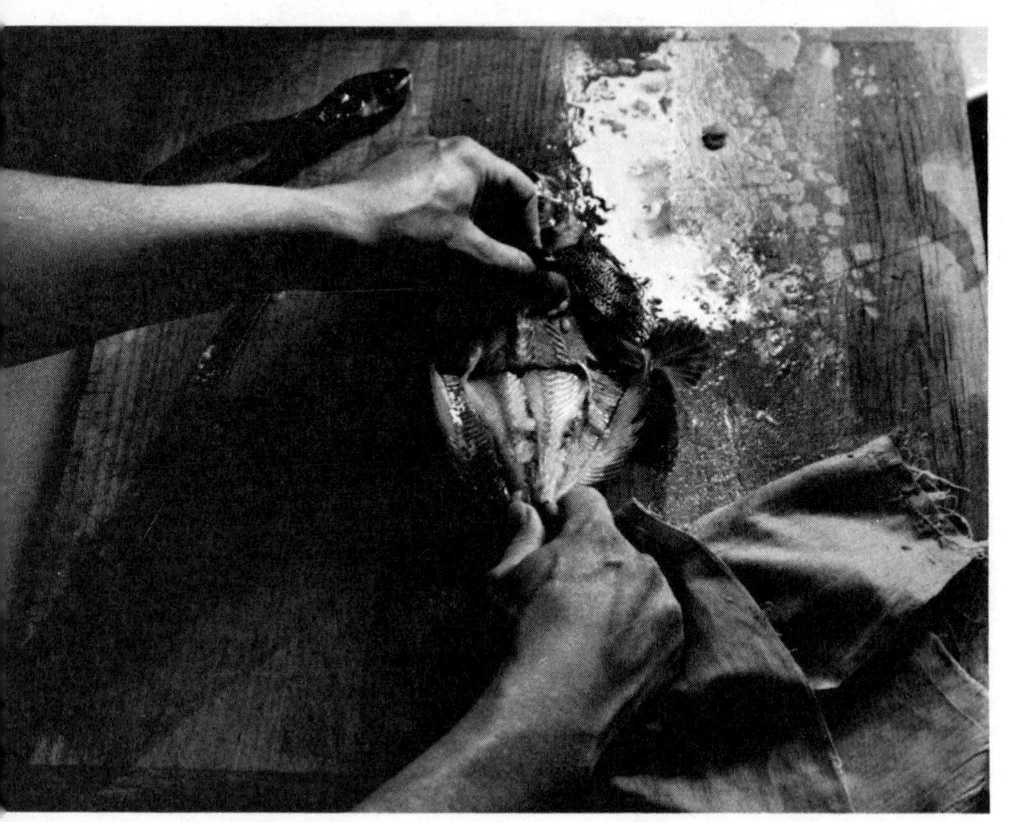

Fig. 4. Grasp the skin and, while holding the tail, pull it up and toward the head.

Fig. 5. Use the tip of the knife blade to scrape the meat away from the backbone to produce the fillet.

in maintaining the integrity of the meat. Without its skin the fish falls apart when it is poached or boiled. The outer membrane of the fish also protects the flesh from drying out when it is baked or broiled. The covering helps to keep the meat moist and seals in the flavors.

KEEPING FRESH FISH

If you know that you will be eating the fish in a day or two, you can store it in the refrigerator. Pat the fish dry and put it in a shallow, covered pan. The body fluids that will seep from the fish as the connective tissue breaks down should be drained off periodically. Another method is to lay the fish on crushed ice before putting it in the refrigerator, but some provision must be made to allow the melting water to drain to avoid soaking the fish and losing both flavor and nutriments. The best temperature for storing fresh fish is between 35° and 40°. This temperature range will keep the fish nicely if it was in good condition when it was refrigerated.

FREEZING FISH

For longer storage, fish should be frozen. Keep the pieces as large as possible. Several small pieces will deteriorate faster than one large one because more surface area is exposed. In selecting the method of freezing, consider how you will eventually use the fish. For example, small fillets that are to be eaten singly can be cut from the fresh fish and wrapped individually before freezing, while a larger piece that is to be baked should be frozen whole.

Preventing Deterioration by Freezing

Three main types of deterioration are arrested by freezing. The first is protein degradation which occurs when protein in fish tissue is broken down by autolytic and bacterial enzymes. Freezing, however, kills only about sixty to ninety percent of the bacteria. The rest are inactivated, but they resume releasing their digestive enzymes as soon as the fish is thawed.

The second type of deterioration, oxidative rancidity, occurs when the unsaturated fats in the fish come in contact with air. For this reason the maximum storage time for oily fish is shorter than that for lean fish. Oxidation affects the taste of the meat. The rate of rancidity increases in warm temperatures, bright light, air, and in the presence of metal ions from copper and iron.

Oily Fish	**Lean Fish**
Chub, Herring, Lake Trout, Mackerel, Pompano, Rainbow Trout, Sablefish (Black Cod), Salmon, Shad, Smelt (salt-water), Tuna (Albacore, Bluefin, Little), Whitefish	Carp, Catfish, Cod, Flounder, Grouper, Hake, Halibut, Ling Cod, Mullet, Ocean Perch, Pickerel, Red Snapper, Sea Bass, Sea Trout, Smelt (fresh-water), Yellow Perch, Yellow Pike

Fish with high oil content become rancid in the freezer after three months, while those with a more moderate oil content will last for five to eight months. Lean fish will be good up to a year. Of course any of these fish can be eaten after longer storage periods, but the quality will not be as high as in the ones consumed within the recommended time.

The third type of deterioration, in which the fish turn brown on the surface, is called the maillard reaction and is most prevalent in white-fish fillets. This process is a nonenzymatic reaction in which the amino acids in the muscle tissue combine with reducing sugars. The most effective way to slow the reaction is to soak the fish in a 0.1-percent solution of ascorbic acid for one to two minutes before freezing them.

Ascorbic acid (vitamin C) is available in powder form from drugstores. Depending on the brand, each ounce of powder will contain a specific number of grams of ascorbic acid. Since a gallon contains 3,785 grams of water, 3.78 grams of ascorbic acid are necessary to produce a 0.1-percent solution. Therefore, stir into each gallon of water the number of ounces of powder that contain about 3.8 grams of ascorbic acid.

For example, one brand of ascorbic acid contains 5,000 milligrams (5 grams) of ascorbic acid per teaspoon of powder. Since 3.8 is 76 percent of 5, use about three-quarters of a teaspoon of this particular powder in a gallon of water to make a 0.1-percent solution. Other brands may have more or less ascorbic acid per unit of powder.

An alternative to soaking the fillets in ascorbic acid is to dip them in a light brine. Soak large fillets for twenty seconds in a gallon of water to which 2 cups of salt has been added and dissolved. Soak smaller fillets in 1½ cups of salt per gallon of water for the same amount of time. The dip not only washes the fillets but also reduces the amount of fluid that is released as they thaw.

In addition to the chemical problem of rancidity, another problem is dehydration. Frozen fish that go through temperature changes (such as those near a freezer door) freeze, thaw, then freeze again. Each time this happens, ice crystals draw moisture from the tissue until it turns the yellow color associated with freezer burn.

The higher the air temperature in the freezer, the worse this problem becomes. Warm air holds more moisture than cold air and consequently causes more dehydration. Zero degrees is the maximum temperature recommended for extended storage of fish products.

Air also acts as an insulator, increasing the time required to freeze a piece of fish. Therefore, exclude as much air as possible from packages before placing them in the freezer. When food is frozen quickly, the frozen water molecules remain nearly the same size as liquid water molecules, but when the freezing process is long, the ice crystals become large, causing greater damage to the tissue. This damage leads to an increase in the amount of "drip," the water that seeps from the tissues as the fish thaws.

A long period of time between when the fish is killed and frozen also contributes to large ice crystals because the fluids lost from the tissue as it sits freeze into crystals independent of the water in the cells. Remember that any frozen fish will suffer some deterioration no matter how it is frozen.

To freeze many small fish, put them in empty milk cartons or in covered plastic containers head to tail, fill the cartons with water, and put them in the freezer. To freeze the fish faster, fill the containers with ice water. Label the package with a waterproof felt-tip pen to designate the type of fish and the date they were frozen.

To thaw the fish, simply place the cartons in a pan or in the sink at room temperature. Soon the blocks of ice can be removed from the containers and the water allowed to drain. The ice keeps the fish cool while they thaw and prevents microbial action.

Often it is more convenient to wrap each fish individually before freezing. This makes them easier to handle once they are solid. One method of packaging the fish is to wrap them in plastic wrap individually while they are still wet from the washing. The moisture will cause the plastic to adhere to the skin, reducing the amount of air that is in contact with the fish. Rubber bands, string, or freezer tape should be wrapped around the package to secure the plastic. To protect the plastic from tears and abrasions, and to reduce the risk of bruising the fish, wrap the whole package in a layer of moistureproof butcher paper or aluminum foil. Because the plastic and the foil are airtight and watertight, they will prevent dehydration. The glazing formed by the water will also prevent air from desiccating and oxidizing the fish. Write the freezing date, type of fish, number of portions, and the estimated expiration date on the outside of the package.

Wrapping the fish and simply dropping them in the freezer is not the best way to freeze them. The freezer should be at its coldest setting at least two hours before the fish are to be frozen. If possible, place them directly on the freezing coils. Otherwise, place them on racks (refrigerator racks work well) so that cold air can circulate freely around them. Avoid overlapping the fish; air should flow around each one.

Fish will freeze slowly, if at all, when they are surrounded by too many packages. For this reason, large quantities of fish cannot be frozen at one time. Most home freezers can freeze two to three pounds of fish per cubic foot of freezer capacity in twelve to fourteen hours. The standard holding temperature for long-term storage is 0°.

Glazing

Commercial fish processors use a method of freezing called glazing. It involves less handling of the fish, and it can be done in large batches. Home freezers can closely approximate the results of glazing if a few easy steps are followed before freezing. The system works equally well on whole fish or individual pieces. After the fish are frozen, they can be individually wrapped or put in plastic bags.

The first step in glazing is to freeze the cleaned fish until they are solid. The time necessary for freezing varies according to the thickness of the fish, the type of freezer, and the number of other packages in the freezer. Fill a small tub with enough cold water to completely cover the fish and add several trays of ice cubes. Wait two or three minutes for the ice to melt. To avoid chunks in the glaze, remove any ice cubes that do not melt. Stir in six tablespoons of cornstarch (brown sugar is an acceptable substitute) in each gallon of water to make a five-percent solution. Add enough ascorbic acid to make a 0.1-percent solution and dip each fish in the water.

Make sure the water covers the entire fish, both inside and out. Remove the fish from the water and let the water freeze on the fish. After each layer of glazing freezes, repeat the process until the glaze is one-eighth to one-quarter inch thick. The cornstarch in the water prevents the glaze from cracking or chipping, and the ascorbic acid helps to prevent discoloration of the fish. Fish that are stored for more than six months should be reglazed. The second glazing goes right over the first, but keep the fish at 0°.

Fish are sometimes frozen in a brine glaze in commercial cold storages. Oily fish don't respond well to this treatment, however, because the salt acts as a catalyst in oxidizing unsaturated fats.

THAWING FROZEN FISH

Fish fillets are best thawed in the refrigerator at a temperature of 40° to 45°. One pound of fish will thaw in twelve to sixteen hours, the time varying with the thickness of the fillet. If fish are thawed at room temperature, surface spoilage can occur unless they are insulated in several layers of newspaper. Thin parts of the fish spoil first because they thaw more quickly than the thicker parts.

If you have a shortage of refrigerator space, wrap the fish in a thick layer of newspaper and thaw them at room temperature. The insulation keeps the surface temperature of the fish nearly the same as the inside temperature, equalizing the rate of thawing.

Cook frozen fish fillets without thawing them to reduce the amount of drip (and the nutrients it contains). Allow additional cooking time for frozen fish. If you want to cook pieces without thawing the whole fish, cut steaks from frozen, dressed fish with a hacksaw or bone saw.

To thaw large batches of fish, or to thaw small batches more quickly, put the fish under cold running water. The length of time necessary for thawing is determined by the weight of the fish and by the volume and temperature of the water and how fast it is running. The average thawing time is two hours per pound. Water that is much above 45° tends to soften the fish, making it mushy. Soaking it too long removes the flavorful juices and oils and reduces the nutritional quality.

Warm temperatures are conducive to bacterial growth and lead to surface spoilage, so if *Clostridium botulinum* spores are present in fish that are thawed in an airtight package, the spores may grow and produce the toxins that cause botulism. Cool thawing temperatures help to prevent these bacteria from developing.

2

Smoking

Smoking, whether for its distinctive flavor or for the practical value of its keeping qualities, is a popular method of processing fish. Many types of fish respond well to smoking, particularly the salmon species and steelhead that are native to the Pacific Northwest and Alaska, and the freshwater fish—trout, crappies, and perch—from the inland sections of the country. White fish from the ocean—cod, haddock, tuna, and sturgeon—and fish that usually are not eaten fresh, such as whiting, shad, buffalo fish, butterfish, and eel, are delicious smoked.

There are two main methods of smoking fish: The first is the cold-smoking process, which produces a lox-type product. Cold-smoking is done at a temperature of about 70° for a relatively long period of time and results in a long-keeping product that is partially dehydrated. The second is the hot-smoking process, which yields a flaky, kipper-type fish. Hot-smoking is the process of smoking fish at temperatures above 70°. The fish are partially cooked as they are smoked and are therefore more perishable than those that are cold-smoked. The steps outlined in this chapter will serve as a guide in developing recipes and formulas that will work best with the particular species and sizes of fish available in any given locale.

Fish-smoking is an art that is acquired through experience. Ultimately you will devise your own formulas and learn to control the smoking process to produce the type of fish product you want. When testing a new technique or when using a new kind of fish, begin by smoking small batches instead of a large batch. And always use the highest quality fish for smoking. The quality of the fish going into the smokehouse determines the quality of the smoked product. Poor quality fish will never improve with smoking.

THE COLD-SMOKING PROCESS

In cold-smoking salmon and steelhead (sea-run rainbow trout), use large, thick-sided fish. Small fish (under ten pounds live weight) tend to be too thin when filleted and become overly dry when smoked. The percentage of meat trimmed away is also rather large on smaller fish. For these, see the section "Smelts and Other Small Fish" in this chapter. Red-fleshed salmon are preferred for cold-smoking, while white-fleshed kings (chinook) are used for kippering. Use fish that have been frozen for at least eight days. The reason for this amount of freezing time is to break down the cell structure of the tissues to guarantee a rapid, uniform absorption of the salt cure.

Preparing the Fish

The first step in the cold-smoking process is to remove the fins from the headless fish (see chapter 1 on how to clean salmon and other round-shaped fish). The easiest way is to grasp the tail of the fish (wear cotton gloves, or hold the fish with a towel), holding it belly down on the table and exposing the two dorsal fins on the back. Place that part of the knife blade closest to the handle behind the first dorsal fin and "push" it off. (See Fig. 6.) Repeat the procedure with the second dorsal fin. Make sure that you push rather than slice.

Flip the fish over and remove the pelvic fins in the same way. These are more difficult to remove. Grasp the tail, tip the fish to the left, and with the knife tip facing down, flip the left belly fin off with a quick backhand motion. Remove the right belly fin (the ventral fin) by placing the knife blade, tip up, behind the fin and pushing. (See Fig. 7.)

To remove the two pectoral fins, lay the fish on its side, lift the fins with one hand, and cut from either the front or the tail-

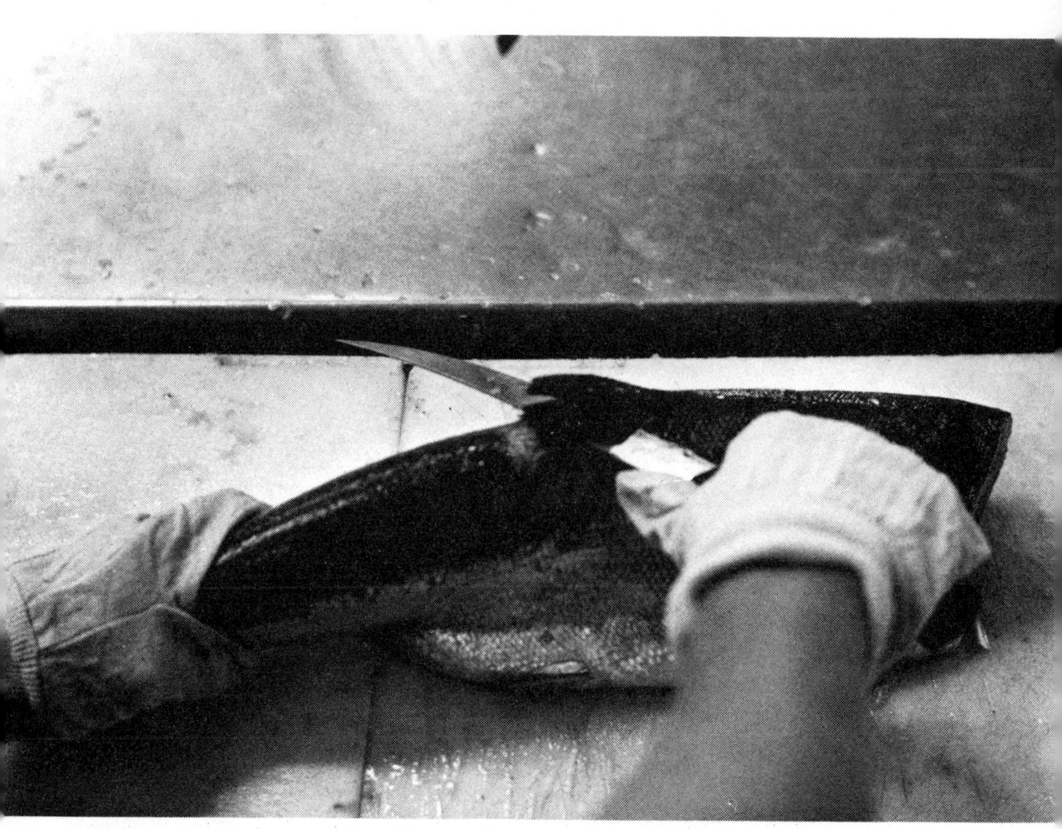

Fig. 6. To cut off the dorsal fin, place the knife blade against the fin and push forward.

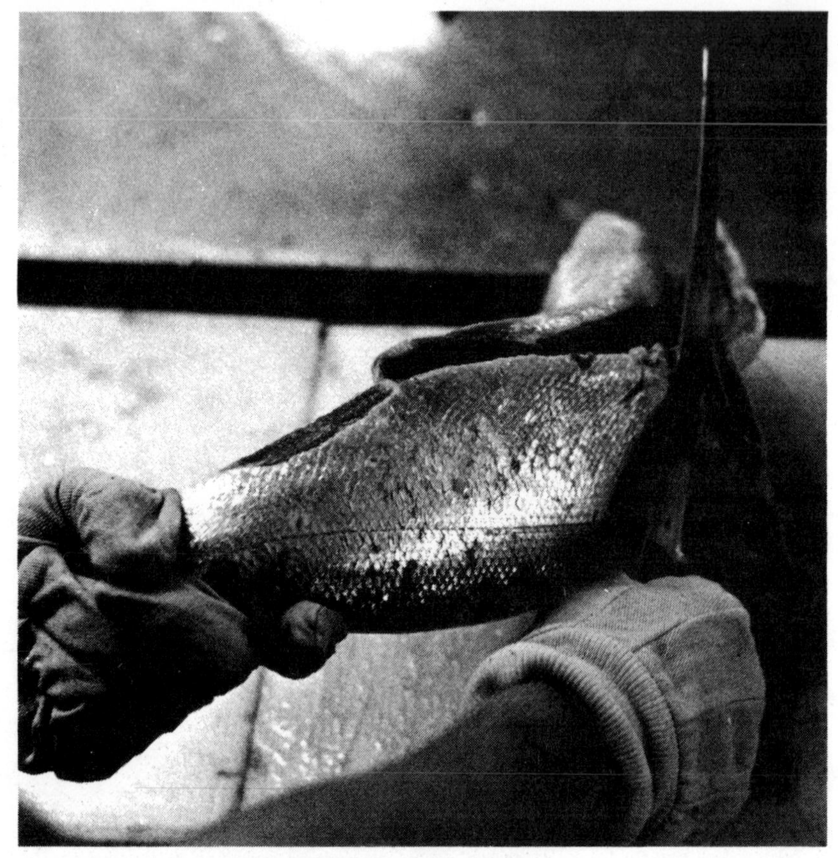

Fig. 7. Remove the ventral fin (the belly fin) in the same way as the dorsal fin.

end. With practice, the entire fin-removal process should take no more than thirty seconds.

After the fins are removed, you must score the skin. To provide an opening to allow the salt cure to penetrate, cut the skin with a single-edge razor blade. Hold the corner of the blade between your thumb and forefinger so that one-quarter to three-eighths of an inch of the blade penetrates the fish. (See Fig. 8.) Beginning at the tail end, make a one-inch slit one-quarter to three-eighths inch deep. Leave a half-inch space, then make another slit, until you come to the head end.

Score in rows one-half to three-quarters of an inch apart and concentrate them in the middle of the upper half of the fish. Do not score too near the top (the back or dorsal side) of the fish or the cuts will interfere with the splitting, and do not place them too near the belly, where the meat is thinner and scoring unnecessary.

Scoring is essential for thick fillets because it allows salt penetration, which keeps the fish from souring in the smokehouse.

Splitting the Fillets

Splitting, or cutting, the fillets is the next step in preparing the fish for smoking. Make a hook to hold the head end of the fish by driving a nail through a ½- by 12- by 30-inch board at an angle that is great enough to catch the fish at the collar. The nail will prevent the fish from slipping. Place it on the nail board so that the head faces to the right with the belly toward you. Make two small cuts, one above the backbone and one below it, from the anal vent to the tail, taking care not to insert the knife more than halfway to the back of the fish. (See Figs. 9 and 10.) This

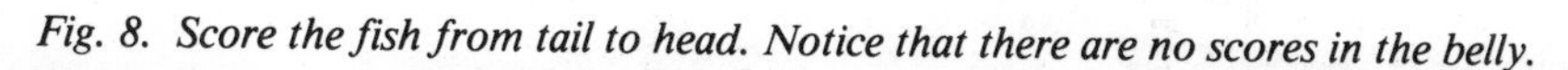

Fig. 8. Score the fish from tail to head. Notice that there are no scores in the belly.

step is called *undercutting,* and it helps reduce the amount of meat left on the backbone after the fillets are cut.

Begin the split by cutting into the neck of the fish just above the backbone. Slice downward until you reach the bone. (See Fig. 11.) Reduce the angle of the knife blade so that it will travel almost parallel to the backbone, cutting through the ribs. After the knife has traveled about three inches along the backbone, put your hand on top of the fish near the neck to hold it in place. Don't use a sawing motion to fillet; draw the knife along in a smooth arc. When the knife comes out at the tail, remove the fillet to expose the backbone. If the splitting has been properly done, a row of small, round "buttons" will be visible where the ribs were cut

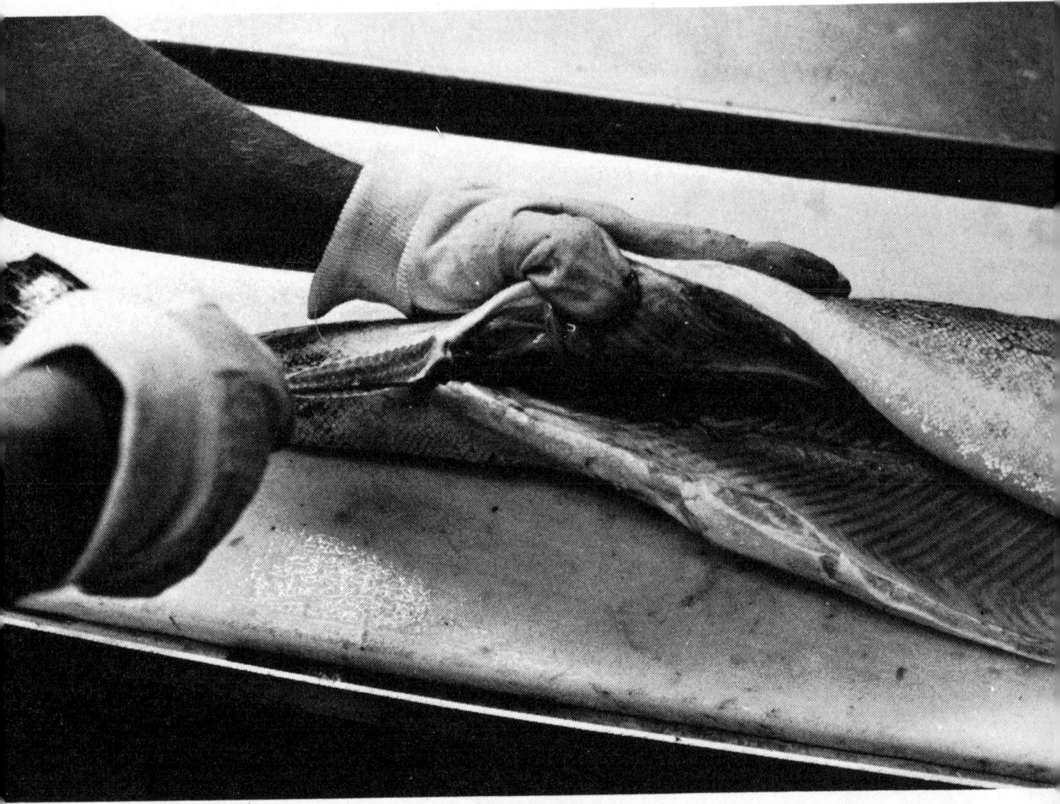

Fig. 9. To undercut the fish, hold the body cavity open and begin the cut on the top of the backbone. Note that the cut begins at the vent.

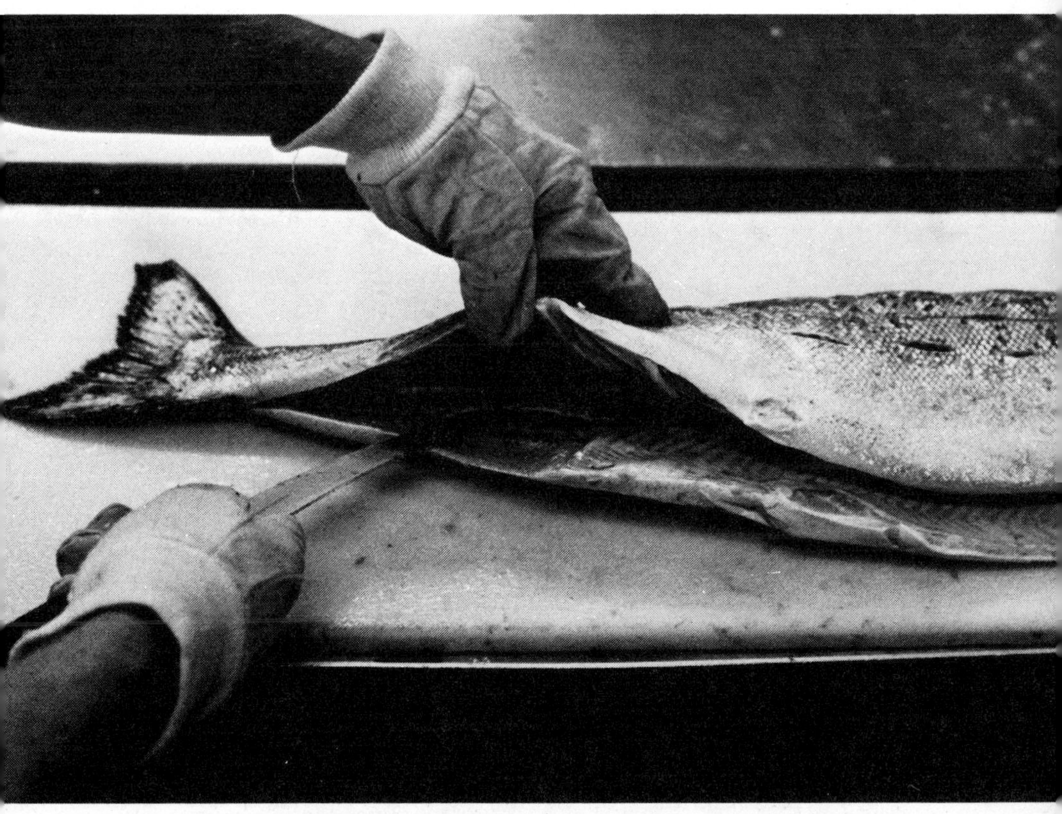

Fig. 10. Make the second undercut beneath the backbone. Be careful not to pierce the skin on the dorsal side, and don't extend the cut out to the tail.

through. Remove the fillet by supporting it along its entire length to avoid cracking the meat. Any separations will be magnified as the meat dries in the smokehouse.

Splitting the left side of the fish is simply a matter of placing the knife under the backbone (the fish remains on the nail with the head to the right), and with a slight upward tilt of the blade run the knife out to the tail. (See Fig. 12.) You now have two fillets that are split, scored, and ready to be boned.

In even the best job of splitting some meat will be left on the backbone. This can be scraped off with a spoon or a knife and fried in butter, formed into a small salmon loaf, or even brined and smoked to make salmon jerky. (See chapter 8.)

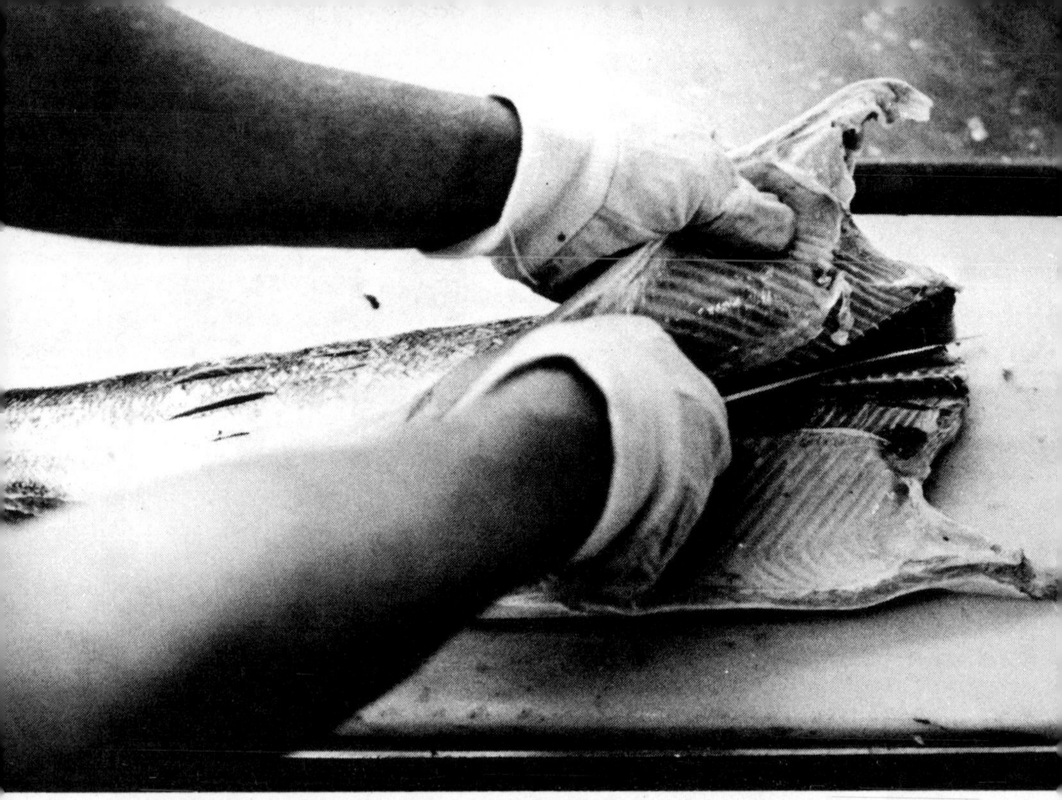

Fig. 11. To split the fish, part it at the neck and make the first cut on top of the backbone. Angle the knife downward to stay close to the bone.

Removing the ribs from the fish requires a considerable amount of practice to be able to do it quickly without losing a large amount of meat. But even a beginner can do a fairly good job the first time. The benefit is that it produces a visually appealing product that is easy to slice when smoked.

Begin by placing the right fillet on the table, skin side down, with the head end to the left. Slip the knife under the bones in the upper left corner and slide the knife to the right, releasing the tops of the bones all the way across. (See Fig. 13.) Repeat this process, beginning at the head, holding the free edge of the bones and their membrane against the meat with one hand. The bones should come free in one piece attached by a thin membrane. Treat the left fillet the same as the right one, but with the head facing toward the right. Make several smaller strokes instead of two or three long ones.

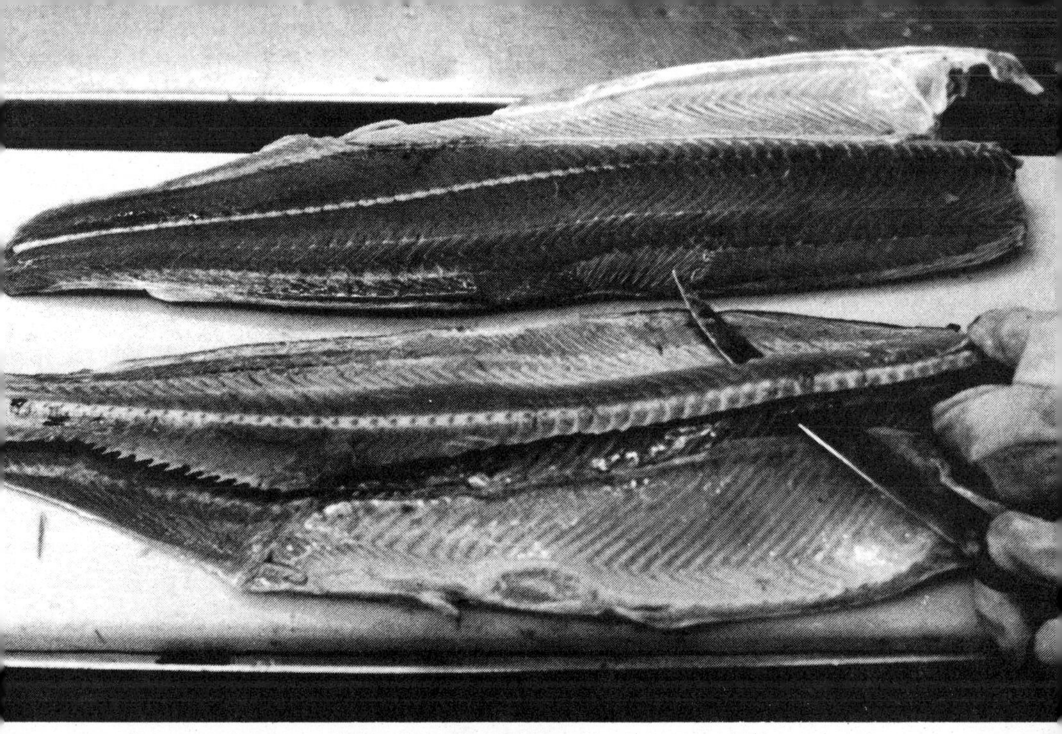

Fig. 12. Make the second cut from the underside of the backbone. Keep the backbone alongside the fish to avoid cutting it off.

Fig. 13. Remove the belly bones, or ribs. Try not to cut too much meat along the bones.

Curing the Fish

The cure for cold-smoked salmon is a half-and-half mixture of medium-grain table salt and brown sugar. It must be well mixed, either by hand or in a blender, so that it is free of lumps. The salt inhibits bacterial growth, while the sugar gives flavor and color to the fish and takes some of the bitterness out of the salt taste.

Lay the fillets on the table, skin side up, with the head pointing to the left. Keep the fish from moving with one hand and push the tail toward the head with your other hand to open the scores. (See Fig. 14.) Sprinkle on the cure liberally and rub it deep into the scores with the palm of your hand. Make sure that the salt penetrates into the flesh.

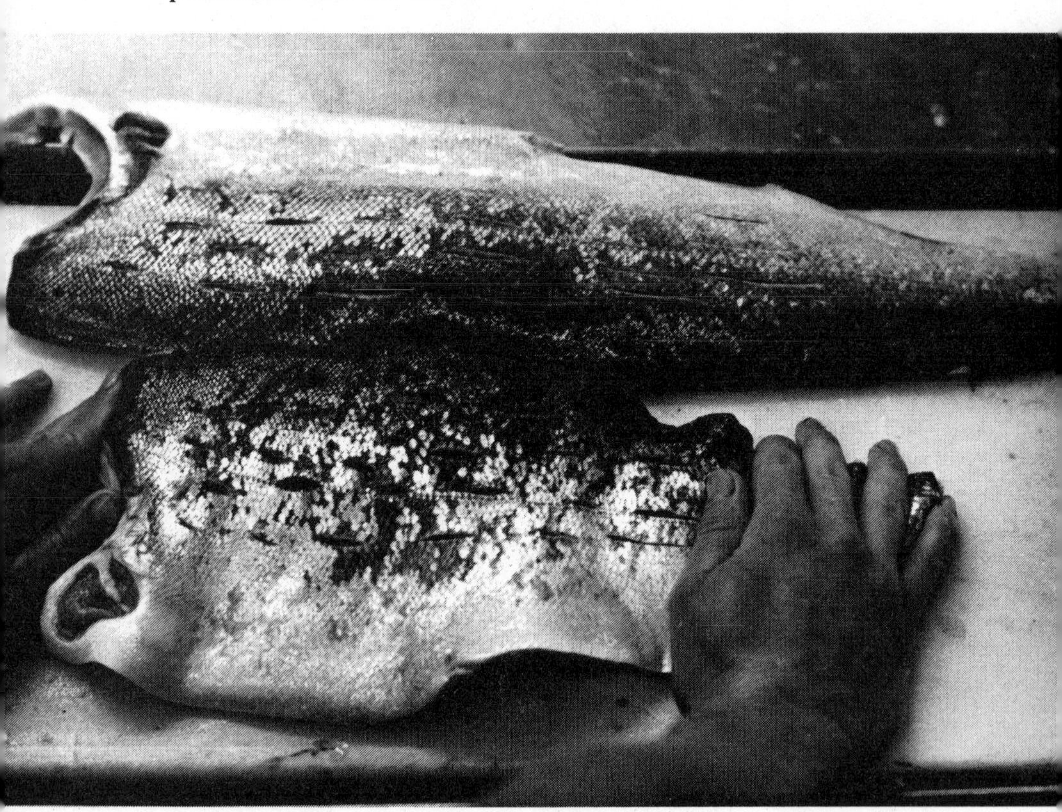

Fig. 14. Bunch up the salmon fillet to open the scores. The open scores allow the cure to penetrate. Apply the cure from tail to head.

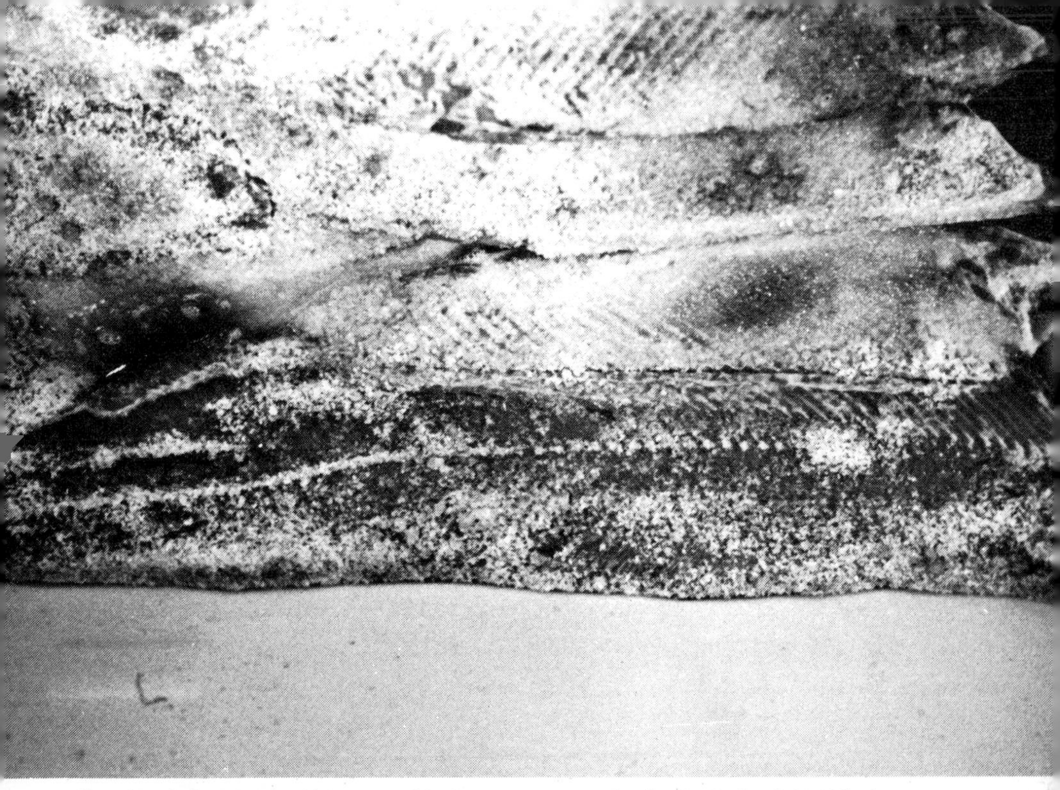

Fig. 15. Salmon sides being cured in brown sugar and salt. Pack the fish with the flesh facing up, as shown.

Spread a thin layer of cure on the bottom of a plastic tub, or on a tray if there are only a few fillets. Lay the fish on it skin side down. Place the backs of the fish to the outside of the tub and the heads toward the ends, allowing the tails to overlap in the middle. This procedure produces a flatter pack. After completing one layer, sprinkle the curing mixture over the tops of the fish until all the meat is covered to a depth of about one-quarter inch. Smooth it by hand to a uniform thickness. Layer the fish four sides high if the tub is deep enough, and smooth the cure over each layer. During the curing process, fluids will seep from the fish and collect in the bottom of the tub. (See Fig. 15.) Do not discard these juices.

Cure the fish for three days at a temperature of 40°. At the end of that time they will be stiff and somewhat shrunken from the action of the salt. Rinse each side individually to remove excess salt and sugar. Rinse the tub and put the fish back in it in the same order. Try to approximate the original placement of the fish to keep from making any new bends in the meat.

Making the Brine

Make a strong brine (nearly saturated or 100 percent) by dissolving approximately three pounds of salt in a gallon of cold water. Reduce the strength to 40-percent by mixing four parts of saturated brine with six parts of fresh water. Use a hydrometer or salinometer to determine the concentration of the brine. These instruments, which look like large thermometers, can be obtained from scientific-instrument suppliers or dairy equipment dealers.

Pour the 40-percent brine over the fish, making sure that all surfaces are covered. Covering the fish keeps it from drying out and prevents bacterial growth. It is common for the top fish to float in the brine. If this occurs, submerge them by placing a clean, weighted basket, screen, or plate on top of them. Then put the fish back in the refrigerator at 40° for two more days. Agitate them once a day to ensure uniform brining.

On the morning of the sixth day, drain off the brine and soak the fish in a tub of cold water for six hours. If possible, the tub should have a water inlet at the bottom and an outlet at the top to guarantee fresh water circulation throughout. Gently agitate the fish every two hours to ensure uniform soaking. After the fish have soaked, take them from the water and put them in the smokehouse. The fish should be fully hydrated, pale, and somewhat soft. Handle the meat gently to avoid breaking it.

Hanging and Drying the Fish

There are many methods for hanging fish for smoking. The type of smokehouse usually determines how the fish are hung. One way to suspend them is on hook or "tenter" sticks, one by two inches, with rows of double hooks down each one-inch side. The fish are hung by their tails on both sides of the stick (one fish per double hook) to form two rows of fish per stick. If hooks aren't available, pierce two holes in each tail and tie two sides together. Then drape them over the stick with the meat sides facing outward. Or cut half-inch notches on both sides of the fillet near the tail. Wrap string around the tail and tie it off before attaching the fish to the stick. (See Fig. 16.) These two techniques will work only with fish that are not skinned because the meat alone is not strong enough to support the weight of the fish.

Place a screen under each stick to catch the fish if any fall off. Hanging fish is good from a sanitary standpoint because the oil that drips from them during smoking falls to the ground instead of puddling up on the sides. The oil puddles would leave raw spots on the fish and serve as breeding grounds for bacteria.

Another method is to arrange several sides on two thin steel rods by piercing the skin in two places near the tail and threading the rods through the fish so that they are two to three inches apart. The ends of the stick are then placed on supports in the smokehouse. For heavy fish, run the rods just beneath the collar at the head, where there is more support.

Fish can also be placed on racks rather than sticks. The racks should be of stainless steel if possible, but chrome-plated racks, refrigerator racks, or racks made from expanded metal also work well.

Place the fish on the rack, skin side down, in a head-to-tail pattern. This maximizes the space on the rack. If you are hanging large fish or using a long rack, support the screen in the middle to keep it from sagging. Any stick will make an acceptable brace. Remember that raw spots will occur wherever the support touches the fish or where it prevents air from circulating around the meat.

Once the fish are racked or hung, they are ready to be smoked. The first step is to dry the fish, either with a fan in the smokehouse or in a shady spot with good air circulation. The fish are dried until a smooth skin, or pellicle, forms on the flesh. The pellicle becomes shiny and dark after smoking. If the fish have been dried outside, put them in the smokehouse for three days at 70°. The dehydration process continues during the time the fish are in the smoke-

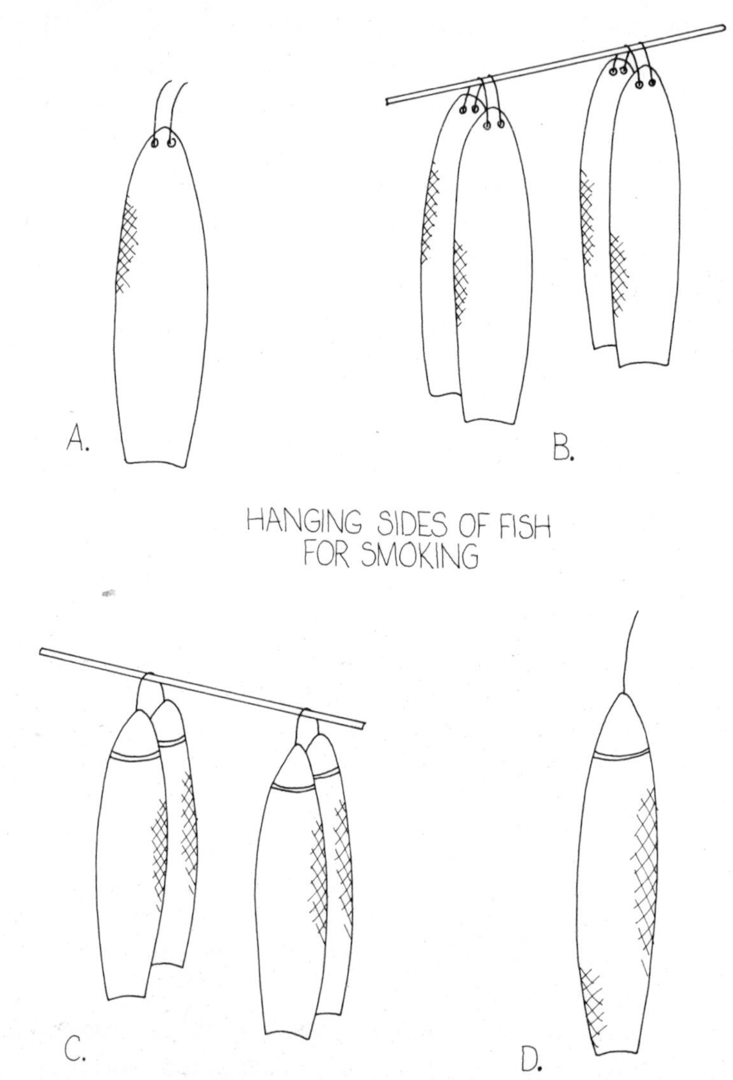

Fig. 16. Part A of this figure is a detailed view of the fish hanging in part B. Two holes have been pierced near the tail end of the fillets and strung with twine so that two sides of fish can be suspended over a stick. Do not place the holes too near the tail because the weight of the fish will cause the string to tear through the meat, letting the fish fall off the stick.

Part D is a similar detail of the fish hanging in part C. Cut a notch on each side of the fillet and wrap a string around the tail end so that the cord catches in the notches. Lead the end of the string over a stick and tie another fillet on in the same way. The cuts must be deep enough to allow the string to grip the sides, but not so deep that the meat will tear.

house. In fact, it accounts for the greatest percentage of the preservative quality of smoking. If you put more than one batch in the smokehouse at the same time, label them with the date and time they were put in.

After smoking each batch of fish, wash the racks in hot soapy water to remove the fish oil and to kill any bacteria. Rinse the racks and sticks and dry them thoroughly before using them again.

Checking the Rate of Smoking

The thickness of the sides, the density of the smoke, the amount of air circulation, and the relative humidity in the smokehouse determine the length of time the fish are smoked. The rate of smoking is directly related to the relative humidity in the smokehouse and the wetness of the fish. High atmospheric humidity and moist fish allow more smoke particles to cling to the fish. If there are "dead spots" or air pockets in the smokehouse, you will have to shift the fish periodically to provide uniform smoke exposure to all the sides. Inspect the fish frequently for best results.

To check the progress of the smoking, cut small slices of meat from the neck of the fish after the first day. If the fish is completely smoked, the flesh will be firm yet resilient and uniform in color and texture. The tail end will be firmer and saltier than the head end because it is thinner and has dried more.

Part of the fish's color and part of the preservative action of smoking comes from particles and vapors of the smoke. When wood is heated without burning, it undergoes the process of destructive distillation. The components of the smoke are formaldehyde and other aldehydes, acetone, methyl and other alcohols, formic and acetic acids, phenols, tar, and other compounds. Coniferous woods produce a resinous smoke that tastes distinctly like turpentine. These woods give the fish a good color, but the flavor of turpentine is hardly desirable. For this reason it is best to make the smoke by burning wood from deciduous trees. The smoke acts as an antioxidant, preventing the oxidation of fatty oils. The formaldehyde and acids retard bacterial growth. As might be expected, the greatest concentration of smoke deposits occur on the pellicle. However, the diffusion of the smoke continues to penetrate the fish for several days after they are taken from the smokehouse.

Storing Smoked Fish

After the fish are smoked, wrap them individually in plastic wrap or waxed paper and store them in the refrigerator or freezer. At 32°, cold-smoked salmon will still be in excellent condition after six to nine days, and even at 60° it will be edible after a week. In the freezer at 0°, the fish will be good for six or seven months.

As the smoked fish deteriorates, it will begin to fade in color, become soft, and lose its sheen. Microbial growth will appear as a white fuzz or slime, appearing first in the scores. Since the molds that grow on fish do not produce mycotoxins, they can often be removed and the fish still eaten. If the fish is just past its recommended storage time, the mold can be wiped off with a clean cloth dipped in a baking soda–water solution and put back in the smokehouse with dense smoke for six hours. It should then be eaten at once or frozen.

Take considerable care with fish that are tightly wrapped in plastic. This creates an anaerobic condition in which *Clostridium botulinum* spores may grow and produce deadly toxins. Bacteria that don't form spores are killed after an hour or two in dense smoke, but those that do form spores aren't destroyed in significant numbers. Always refrigerate these fish.

SMOKED SALMON STRIPS

To make smoked salmon strips, bone and fillet the salmon as for cold-smoking and cut off the collar at the neck. To do this, lay the fillet on the table with the meat facing down and cut the thick pieces of skin off at an angle so as not to waste too much meat. The sides need not be scored to make strips.

Carefully turn the fish meat side up and cut the first strip from the thick, back part of the fish. Make the cut in one motion from head to tail in a one-inch wide strip. As more strips are cut and you approach the belly, begin to make the strips up to one and a half inches wide. This helps to compensate for the thinness of the meat at the belly. However, if they are cut too wide, the salt will not penetrate deeply enough, and the fish will sour in the smokehouse. If the strips are cut too thin, they will curl and the meat will shrink during the drying process. Separate the strips into belly strips and thick back strips. Soak the belly and back strips separately in solutions of 100-percent brine in a plastic or other noncorrosive container. The thick strips should soak for forty-five to fifty minutes, while the thinner strips should soak twenty-five to thirty minutes. After the strips have soaked, pour off the brine and fill the tub with fresh water. Then immediately pour off this water and turn the strips onto the worktable to drain. Leave the strips on the table for a few minutes to allow the water to drain off. This prevents the water from soaking out the salt.

Hanging and Drying the Strips

The easiest way to hang the strips is to tie two together and drape them over a stick. For a large number of strips, cut 20-inch lengths of string. To do this, wrap the string around a 10-inch card as many times as necessary for the number of strings you require. Then make one cut through all the wraps to get the 20-inch lengths and tie each string in a loop. Now make a slip knot as shown in Figs. 17 and 18 by placing your thumb and forefinger through one end of the loop and laying the long, standing part held in your other hand over the short piece of string between the two fingers. Transfer the loop on the thumb to the forefinger. Put the end of the strip of fish into the double loop and cinch it tight.

As you tie the strips together, stretch them out flat along the edge of the table so that the string is pulled straight between the strips of fish on either end. This keeps the strips and the string from becoming a tangled mess.

Fig. 17. Tie the string in a loop. Turn your right hand over so that the string loops the finger and thumb.

When you are ready to hang the strips on the sticks, simply slip the loops over the stick so that they are evenly spaced to a minimum of ½ inch. The stick should be two inches wide to ensure adequate space between strips on the same string.

Hang the strips in a cool, breezy area until they begin to form a pellicle (usually in three to five hours), or put them in the smokehouse to start drying. Make sure the pellicle forms before beginning the smoking phase. As with any smoked fish, good air circulation is important. Strips that touch each other will stick together, leaving raw spots.

Keep the fish in a 70° smokehouse with a moderate amount of smoke for three days, or until they are firm and dry. For a hard smoke that keeps well, the strips can be left in up to five days. If the fish smoke too long they become very hard and impossible to chew. After they are smoked, the strips can be cut into shorter lengths and wrapped in plastic. If smoked for three days, they will keep for one month in the refrigerator and for eight months in the freezer. As with cold-smoked sides, color loss and softening of texture are the first signs of deterioration.

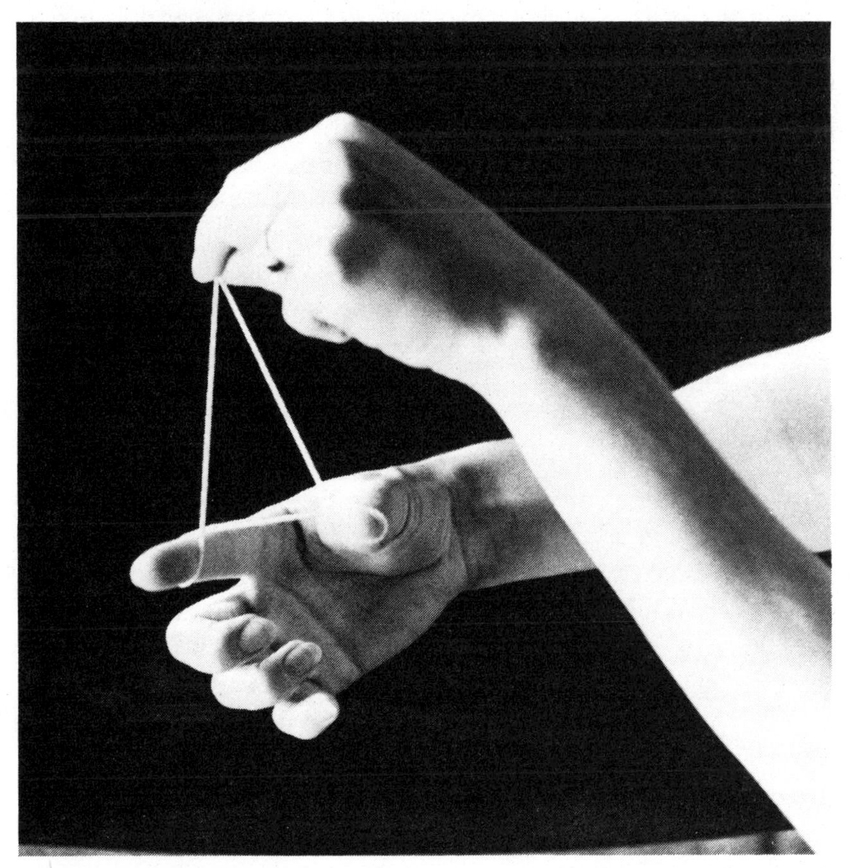

Fig. 18. Move the loop on the thumb over to the forefinger to make a single double-loop to hold the fish.

ANOTHER METHOD FOR SMOKING SALMON

To produce a less perishable product that is somewhat firmer in texture, trim the fins from a headless, dressed salmon and thoroughly wash the body cavity. Score the fish with a single-edge razor blade with just one row of scores. An exceptionally large fish will require two rows. With a knife, trim the tail fin to a length of one inch. If you are processing many fish, you can set up an assembly-line system for each stage in the procedure. In this case, chop off the tail fin with a hatchet.

Next, cut off the belly. Since the belly has the thinnest muscle wall in the fish, it tends to curl as it dries, making it difficult to handle. Salt the bellies separately from the rest of the fish and pickle them later.

The easiest way to cut the bellies is to use a holder so that the fish will be belly up. This makes it easier to remove both sides of the belly in one cut. Make the holder by nailing two 14-inch 1 x 2s three inches apart to a 6- by 14-inch piece of ½-inch plywood. Lay the fish in the holder back side down with the head to the right.

To start the cut, place the knife about one-third of the way down the collar, just at the point where the muscle wall begins to thicken. Push the blade along in a straight line until it exits just ahead of the anal fin. In your other hand, hold the two belly pieces as they are released by the knife.

Remove the fish from the holder and split it from head to the tail, starting on the top of the backbone. Stop the cut before it reaches the last two inches of the fish. Be sure to undercut the fish to avoid wasting meat. Repeat the cut on the underside of the backbone so that the two sides remain connected at the tail, with the free backbone in the middle. Twist the bone until it snaps off.

Dip the fish in a chlorine-bleach solution (one tablespoon chlorine bleach per four gallons water) and place it on a rack to drain. Meanwhile, thoroughly wash a paraffin-sealed wooden-barrel or a plastic pail with soap and water. Rinse the inside with the chlorine solution.

Fill a box 25 by 8 by 8 inches halfway with medium-grain

table salt and dredge the fish in it, allowing as much salt as possible to stick to both the inside and outside of the fish. Be sure to push some salt into the hinge at the tail to prevent bacterial action at that vulnerable spot. Sprinkle a small handful of brown sugar inside the body cavity and add a dash of salt for good measure. Use the following ratios of pounds of salt to pounds of fish:

salt	to	fish
3– 4 lbs.		10 lbs.
6– 8		20
12–16		40
30–40		100

Sprinkle salt into the barrel or bucket until the bottom is covered to a depth of an eighth of an inch. Place the fish so that the tails are to the center, taking care not to spill the salt and sugar from between the two sides of each fish. The first few fish go in with the back to one side; the remainder go in with the backs to the other side. This prevents the fish from becoming too curved at the end of each layer.

Over each layer of fish sprinkle a couple of handsful of salt. Do this also on the top layer. If you have more fish than the container will hold, wait until the next day; the fish will settle and the container will partially fill with brine, leaving room for the rest of the fish.

Make a strong brine (about three pounds of salt per gallon of water) and fill the container. Sit the container in a cool place for three or four weeks. The fish can stay in the brine for six months in cool climates without any appreciable decrease in quality, but it is critical that they remain covered at all times. So long as they are covered with brine they will not turn yellow-rust or acquire the strong taste of salt-burned fish. Air, sunlight, and salt are always a bad combination for storing fish.

Use a hydrometer or salinometer to determine the amount of soaking time necessary—according to the strength of the brine—for the fish when they come out of the barrel. Save part of the brine from each barrel or container to check its salt concentration. The chart below lists the approximate number of hours that the fish should soak in cold, slowly running water for different brine concentrations.

Brine Concentration (Hydrometer Degrees)	Soaking Time (Hours)
75	24–25
80	28–30
85	32–33
90	35–36
95+	40–42

To soak the fish, remove them from the barrel, drape them over 1 by 2 inch sticks, flesh side out and immerse them in the water. After sixty percent of the soaking time has elapsed, raise the tail out of the water to where the anal fin would start if it were still attached. Since the meat is thinner at the tail, it soaks out faster than the rest of the fish. Small pieces should be cut from the meat and tasted periodically to test for saltiness. Remember, as the fish lose moisture in the smokehouse, the salt will become more concentrated in the meat, so a wet fish that tastes just right may be too salty after it is smoked.

When the fish have finished soaking, remove them from the water and hang them in a cool, airy place to dry. Placing the freshened fish in front of a fan will facilitate drying. The fish can also be hung in the smokehouse immediately if you have a fan to circulate air around them.

Leave the fish in the smokehouse for seven to nine days, depending on the amount of humidity in the air and how fast they dry out. Move them around once or twice a day so that they all receive the same amount of smoke. Never let the temperature of the smokehouse go above 70°. The fish will spoil if they are left at too high a temperature for too long a period.

This type of smoking must be done in cool weather. The fish may be frozen until conditions are suitable or they may be salted.

Or try smoking the fish at night and refrigerating them during the day. You can try an air conditioner but it will increase the humidity in the smokehouse.

After they are smoked, the fish have a salt content high enough and a moisture content low enough to keep for long periods at 40° or below. To store them, cut the sides apart at the tail and tightly wrap them in butcher paper or plastic wrap. They can also be skinned, boned, and sliced and placed in small plastic containers. Brush a light coating of vegetable oil on top of the slices, fill the container to the top, and try to remove as much air as possible from the container. "Burp" the air out by squeezing the container before you attach the lid. The fish will keep for several months in the refrigerator and for a year in the freezer.

Because fish shrinks as it dries, the slices are sometimes rather narrow when they are cut vertically. Make wider slices by cutting the fish on an angle. Before slicing mild-cured fish (that are a bit softer than the harder smoked fish), put them in the freezer for about an hour. Semifrozen fish can be cut much thinner and more neatly than cool fish.

COLD-SMOKED SMELT AND OTHER SMALL FISH

Trout, crappie, perch, smelt, and other small fish can be cold-smoked to yield a good product. Larger fish and trout can be

Fig. 19. With a knife, make wider slices from the smoked fillets by cutting the meat on an angle.

dressed as outlined for salmon, or they can be opened, cleaned, and spread apart with toothpicks to dry and smoke. The toothpicks prevent the fish from curling as they dry and allow air and smoke to come in contact with all parts of the fish. Small fish can be cut through the back (without cutting through the skin at the belly) to remove the entrails, then spread open to smoke.

To clean smelt, follow the diagram in Fig. 20. Cut off the head and gills and then the belly (with the organs in it). Make a cut from the head to the tail. Because the bellies are thin, not much meat is actually wasted. Some people merely wash smelt and leave them round with the organs still in them.

Because salt absorbs moisture, whereas a brine solution does not, cure the fish for three to twelve hours in a fifty-fifty brown-sugar–salt mixture to begin draining some of the fluid from the fish. Thick pieces of fish may require several days to cure when they are packed in salt.

To salt the fish, dredge them in the sugar-salt mixture in a small box or pan. Scatter the mix to a depth of one-eighth inch on the bottom of a wooden or plastic tub and put the first layer of fish in it skin side down. Scatter more salt on top and add the next layer. Put in the last layer skin side up and sprinkle it liberally with salt.

The length of time the fish should remain in the sugar-salt mixture depends on whether they are split or round, how fresh they are, and the oil content of the meat. After the fish have been cured, rinse them in fresh water to remove the salt. Taste a small portion of the fish. If it tastes too salty, soak it in cold running water to remove some of the salt. Otherwise, it is ready to be dried and smoked.

Hanging smaller fish is similar to hanging sides of salmon. Hooks and string can be used to secure the fish to sticks, or racks can be used to suspend the fish in the smoke. For small, light fish, make racks from 2 x 2s and half-inch-mesh chicken wire. If you are using a large rack, brace it in the middle. A metal rod or a 1 x 2 board with the one-inch dimension facing up works better than a 2 x 2; the fish won't receive an adequate amount of smoke if they

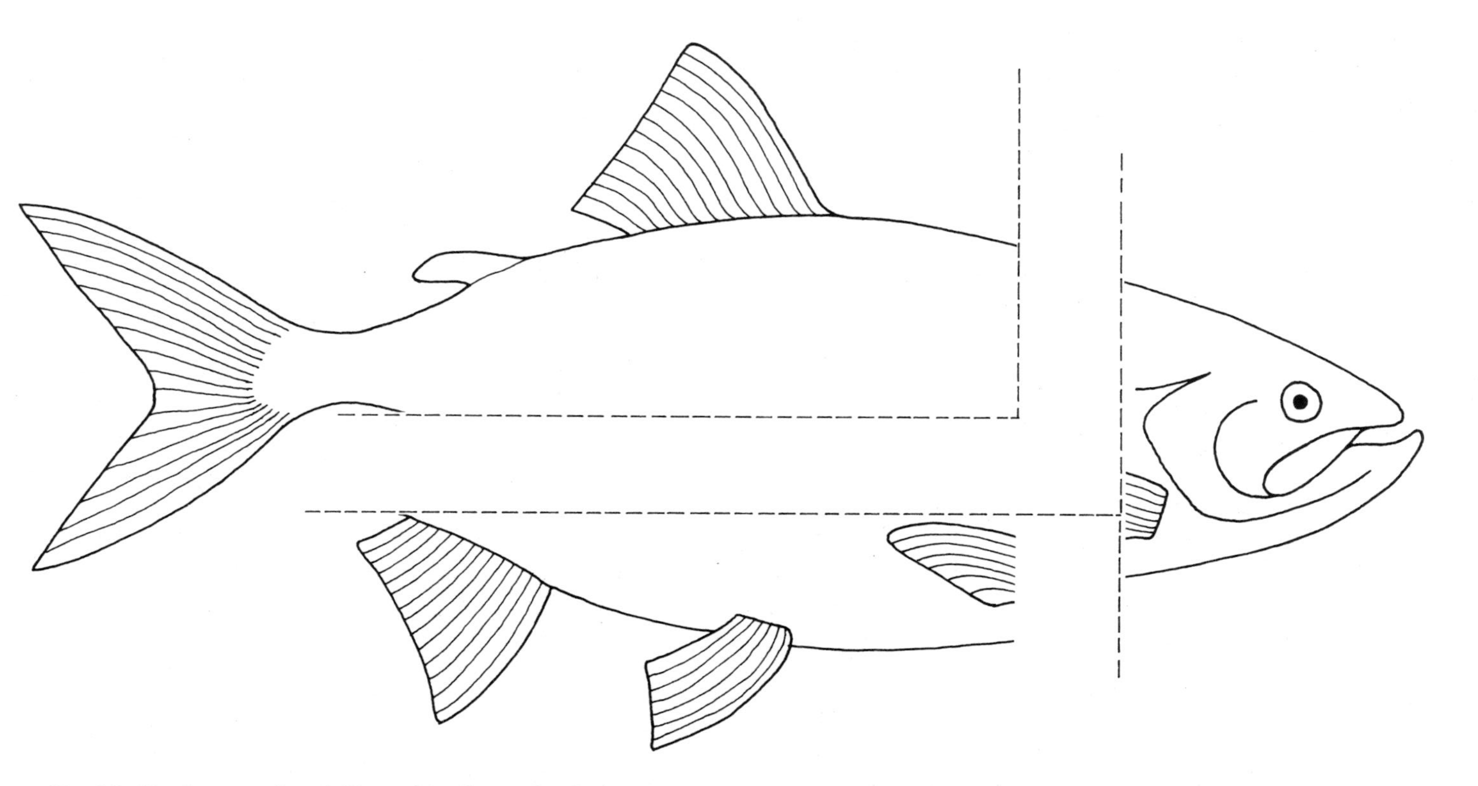

Fig. 20. To clean smelt quickly and easily, make the cuts indicated by the dotted lines. By doing this, you will have little waste. There is no need to remove the viscera.

are laid across the broad two-inch surface. Wrap the chicken wire around the wood once before nailing or stapling to make a stronger bond.

Smaller fish smoke faster than thicker sides and as a result tend to be drier. Depending on the thickness of the fish, one to three days in a 70° smokehouse is adequate.

COLD-SMOKED HERRING

Cold-smoked herring has a variety of names, depending on the type of cure and the length of time in the smokehouse. Reds, silvers, bloaters, and kippers are all smoked at a temperature of 80°. The first three are made from *gibbed* herring—that is, herring with the gills and heart removed. To remove the gills, make small cuts on either side of the throat and pull out the gills and exposed heart, leaving the other organs intact. To cure the herring, put them in a barrel or plastic container and pour in pure uniodized salt and work it into the meat. If enough salt has been added, a brine will form and cover the fish.

Herring goes through a seasonal cycle in which the fat content of the flesh varies. In the late summer and early fall the meat is high in oil. In the spring the fish approach the spawning season and have a lower fat content. Oily fish caught in the fall make the best product. Immature herring are called *sild* and often are packed as sardines.

Keep reds at a cool temperature for one week in the salt mixture. Keep silvers for ten days and bloaters for one day. After the proper amount of time has elapsed, rinse off the salt from reds and bloaters and hang them to be smoked. Soak silvers in slowly running water for twelve to eighteen hours before hanging them.

Smoke silvers and bloaters for twelve to fifteen hours in a thin smoke. It may be necessary to rotate the fish once during the process so they all receive the same smoke exposure.

Reds are smoked for twelve hours every other day for seven days. Keep them cool between smoking sessions. After they are smoked, reds will keep for three or four days if refrigerated.

To prepare kippers, split fresh herring along the back, leaving the belly intact. Clean the fish thoroughly, taking care to scrape the kidney off the backbone. Or remove the backbone altogether. Make a strong (100-percent) brine and soak the fish for twenty to twenty-five minutes. Smoke them for approximately twelve hours at 80°. Before smoking kipper fillets, brine them for ten minutes. Then smoke them for the same length of time and at the same temperature as the whole fish. If they are not used immediately, kippers should be canned or refrigerated. They will keep for four or five days in the refrigerator. Canned fish do not require refrigeration.

COLD-SMOKING WHITE FISH

Cod, halibut, and other ocean white-fleshed fish are delicious when they are cold-smoked before being cooked. Smoked haddock is called *finnan haddie.* Shark meat can be smoked, although sharks, skates, rays, and other members of the elasmobranch family contain one-percent urea in the tissues and sometimes acquire an ammonialike flavor when dried with heat. Tuna and sturgeon should go through a short smoke before canning.

Black Cod (Sablefish)

To dress a butchered black cod (sablefish), cut off the fins and score the fish as described for preparing salmon. Cut the neck open to expose the body cavity. The first cut to remove the backbone is made along the entire length of the bone, starting at the head. Cut the fish along the top side of the backbone first and then along the underside. The knife should not pierce the skin on the back. At this point the backbone is free of the meat on either side but is still attached along its back.

Make the next cut from the head end of the fish by cutting through the row of bones still connected to the backbone. The cut comes out near the tail, and the result is a fish that opens like a book with the skin as its cover and a detached backbone with a tail. Trim the edges of the belly if it is especially thin.

Soak small black cod (under five pounds) in a 100-percent brine for thirty minutes. Soak larger cod for forty minutes. At the end of the soaking time, remove the fish from the brine and, without rinsing, hang them on hook sticks. The backs should face each other on the same stick. Because black cod sour quickly, it is essential that the fish not touch each other when hung. If two fish on opposite sides of the stick persist in touching, wad up a paper towel or small piece of newspaper and place it between the fish to separate them.

The fish should remain in a 70° smokehouse for eighteen hours. After they are smoked, the fish will be golden. The depth of the color varies with the length of time they remain in the smokehouse. The dried salt from the brine sometimes appears as a white powder on the skin, but it is harmless. Simply wipe it off with a damp cloth. If the fish are to be eaten within three days, they can be stored in the refrigerator at 40°. If they are to be stored for up to six months, they should be frozen at 0°.

The traditional way to prepare smoked white-fleshed fish, whether cod, halibut, haddock, or others, is to poach or steam them. To poach the fish, put them in a skillet and cover them with milk. Simmer them (but never boil) for eight to twelve minutes, or until the fish just begins to flake when touched with a fork. Serve the fish with boiled potatoes.

Another poaching method is to simmer the fish in milk for four minutes. Drain off the milk and save it for a white sauce. Remove any bones and arrange the pieces in a greased casserole. Pour heavy cream or a white sauce over the fish and dust the top with white pepper. Put the casserole in the oven and broil it until the top is golden brown.

White-fleshed fish are also good when steamed over boiling water and then brushed with melted butter.

THE HOT-SMOKING PROCESS

Hot-smoking produces an excellent product, although the fish are more perishable than those that are cold-smoked because they are only partially cooked. Almost any kind of fish responds to hot-smoking, although white king salmon is preferred because of its high fat content. In hot-smoking, the fish are prepared just as in cold-smoking up to the point where they are to be cured. However, in hot-smoking they are not scored. Cut the fish into squares about three inches on a side. Make the thickness and the size of the pieces as uniform as possible to ensure the same rate of brine and smoke penetration in each piece.

Make a brine by mixing two pounds of salt (about 3 cups granulated or 4½ cups flake salt) per gallon of water. The brine should read approximately 75° on a hydrometer. Sugar, spices, and herbs can be added to the brine to flavor the fish. Experiment with spices such as white pepper, allspice, cloves, mace, black peppercorns, garlic, ginger, onion powder, and dried mustard, and herbs such as bay leaves, dill, and oregano. Also try chili peppers, maple flavoring, or molasses. Depending on the amount of cure desired, the thickness of the meat, and the oil content, soak the fish for two to four hours in a cool place. Some people add food dye to the brine to color the fish, but it isn't necessary.

At the end of the soaking time, rinse the fish thoroughly in fresh water to remove all traces of brine. Gently press the fish to squeeze out excess water and lay the pieces, skin side down, on racks that have been brushed with vegetable oil. Place the racks in a cool spot, out of the sun, where a breeze or fan can dry them. After a few hours a pellicle will begin to form and the fish will be ready for the smokehouse.

When the fish are first put into the smokehouse, leave the front door open a few inches for a while to allow water vapor from the fish to escape. This is especially important when the fish haven't had a chance to dry before being put in the smokehouse. As the fish begin to dry the door may be closed as the exhaust vent will be adequate for removing the moisture.

For the first eight hours in the smokehouse the temperature must be maintained at 90° to continue the drying process. After four hours, smother the fire to increase the smoke as much as possible without raising the temperature. Close the damper to reduce the air flow and close the exhaust opening in the smokehouse. At the end of the eight hours, raise the temperature to 140° and smoke the fish for another two hours. This is a relatively long process, so plan your batches for the smokehouse early in the day. This will leave plenty of time to tend them.

Inspect the fish to determine if they are done. For a drier product, leave the fish in the smokehouse for another hour or two. The fish should have a smooth brown surface and be moist and flaky inside. White curdish chunks on the top are a result of smoking the fish too quickly before they have had a chance to dry thoroughly. You can avoid this by making sure that the pellicle forms before smoking the fish.

Another time-temperature system for kippering is to place the fish in the smokehouse after they have begun to form a pellicle and raise the temperature to 120°. Maintain the temperature for two hours. After the first hour, drops of water will form on the surface as moisture is driven out. Beginning at the third hour, I raise the temperature to 140° and hold it for five hours, or until the fish are firm.

A third system is to keep the fish at 80° for ten to twelve hours and then raise the temperature to 175° for one hour, or 250° for thirty minutes. The high-heat stage of the smoking can be done in the oven.

After smoking the fish, remove them from the smokehouse to cool in the open air or under a fan for several hours. If you put them in the refrigerator at this point, the hot fish will turn to a soggy mush. As they cool the fish can be brushed with vegetable oil to keep them from drying out. Wrap the cooled fish individually and refrigerate them for up to a week or freeze them for longer storage.

Preparing Buckling

Buckling is the British name for hot-smoked herring. To prepare Buckling, use only fresh herring caught in the late summer or fall, when the fish are oiliest. Gib and salt the fish and put them in a plastic container for twelve hours. Rinse them in fresh water and hang them on sticks by running the stick under the gill cover of one fish and through the mouth. Then put the head of the first fish under the gill cover and through the mouth of the second fish. It is sometimes simpler, unless a large number are involved, to hang the fish singly instead of doubling them up. Or they can be laid out on racks.

Air-dry the fish and put them in the smokehouse. Bring the temperature up to 130° and build up a dense smoke. After two hours rearrange the fish so they all receive the same smoke and heat for two more hours. Remove the fish and cool them in the open air before refrigerating them. They will keep for three or four days at a temperature of 40° or below.

Smoked Oysters

Anyone who has ever eaten smoked oysters knows what a delicious snack they make. Small oysters are better than large ones, but large ones can be made smaller by simply cutting them up before smoking them. In Washington, and perhaps in other states, it is illegal to remove an oyster shell from the beach. This is because the life cycle of the oyster spat requires a certain amount of time attached to an empty oyster shell. If the shells from shucked oysters are not returned to the water, the cycle is broken and the oyster spawn will not survive.

To shuck an oyster, hold it in one hand with the flat side of the shell facing up. Force a knife between the shells near the thin end, or the "bill." (See Fig. 21.) Cut the large adductor muscle (the one holding the shells together) on the upper shell close to the shell. (See Fig. 22.) Cut the lower half of the same muscle that is attached to the deep half of the shell and pull the two shell halves apart. Check for bits of broken shell in the oysters. Oysters should be shucked and refrigerated as soon as they are caught to keep them from spoiling.

Smoke the oyster meats as soon as possible. Steam them in a steamer (or in a collander set inside a large pot of boiling water)

Fig. 21. A gloved hand is best for holding an oyster so that you can insert the knife in the "bill."

for five minutes. The steam firms up the oyster meat and enables the smoke to penetrate it more evenly. The steaming also removes the protective mucous coat on the meats, so they must be put in a brine solution at once to prevent rapid oxidation.

Prepare the brine solution by mixing a pound and a half of salt (2 heaping cups) for each gallon of cold water. Stir thoroughly to ensure even mixing. Let the oysters soak in the brine for five minutes before draining them.

Place the oysters on a rack or screen (oiled with vegetable oil) and smoke them at 110° for one hour in thin smoke. Control the ventilation so that the air circulates enough to form a pellicle on the oysters. After the first hour, increase the smoke density and raise the temperature to 125°. Smoke the oysters for another hour, remove them, and cool to room temperature. The smoked oysters then either can be frozen or canned.

Smoked Clams

Clams are excellent when lightly smoked. Keep them alive until they are to be processed by putting them in a barrel or other watertight container full of seawater. Replace the water every few hours to supply adequate oxygen. If possible, keep the clams for several days and feed them oatmeal and cornmeal to clean out their intestinal system. Change the water frequently, and don't let it get too warm. To keep the clams for a few hours, put them in a wet burlap bag. Discard any clams whose shells open before processing; these clams are dead and are likely to be spoiled.

Open the clams by steaming them in the pressure cooker at one atmosphere of pressure with the vent open. Remove the meat from the shell, split the neck, and cut off the stomach. Separate the necks from the rest of the clams, and separate the foot of each clam from the folds of meat that constitute the body wall (the mantle). Soak the clam meats in a brine solution—prepared in the

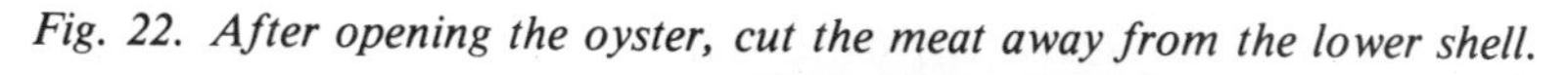

Fig. 22. After opening the oyster, cut the meat away from the lower shell.

same manner as for oysters—for five minutes and drain them thoroughly.

Arrange the meats on an oiled rack and place them in a 105° smokehouse. Smoke the clams for thirty minutes in thin smoke to form a slight pellicle. Build up the smoke and continue smoking for two hours more. After the clams have been in the smokehouse for two and a half hours, raise the temperature to 140° for an additional thirty minutes.

Cool the clams to room temperature and can them or store them in the refrigerator. They will keep for several days.

From time to time in the summer months, microscopic dinoflagellate plankton "bloom" in the waters where clams live. The tiny plants give the water a reddish cast, hence the name "red tide." Clams are filter feeders and in the course of their feeding they pump the plankton through their systems. The toxins produced by the plankton concentrate in the liver of the clams, and they are not denatured during cooking.

Fish and game commissions control the clamming areas that are open during red-tide conditions. Public health commissions in most counties also have information about which beaches are affected and where clams can be safely dug.

3

The Smokehouse

Depending on the type and amount of smoking you will be doing, your smokehouse can be either simple and portable or complex and permanent. For most backyard operations, however, a small, dual-purpose smokehouse is sufficient. Dual purpose means that you can both cold-smoke and hot-smoke fish within the same unit. While it is possible to use many different containers, from barrels or old refrigerators to small, electric, store-bought fish smokers, the smokehouse described in this chapter is constructed of lumber and exterior-grade plywood. To make the smokehouse portable rather than permanent, simply bolt it together so that you can disassemble it and put it away.

Constructing the smokehouse is just as much an art as the actual smoking of the fish. Whether the smokehouse is plain or fancy is up to you. Just remember that adequate smoke production and plenty of air circulation are the two most important characteristics in a good smokehouse.

BILL OF MATERIALS

1 24″ x 24″ piece of thin-gauge sheet metal, for baffle
1 x 2 lumber for baffle braces
10′–15′ of 6″ stovepipe
2 right angle elbows for 6″ stovepipe
1 6″ damper
6 3″ metal butt hinges
4 hook and eye latches
2 60″ lengths of light chain, or twine, to control the width of the door opening in front
2 handles, for raising and lowering the doors
shingles for roof (optional)
1 x 2 lumber for ledgers and the strips between them:
- 8 46½″
- 4 8″
- 12 14¼″
- 2 9″
- 2 13″

2 x 2 lumber
- 2 47″ on the upper edge of each piece and 46¾″ on the lower edge of each piece
- 1 33″

2 x 4 lumber
- 2 46½″
- 2 41″
- 2 48″
- 1 33″

¾″ exterior plywood
- 2 72″ x 48″
- 2 36″ x 24″
- 1 36″ x 48″
- 1 36″ x 44″
- 1 42″ x 54″

¼″ exterior plywood, for smoke spreader
- 1 46″ x 35½″

BUILDING THE SMOKEHOUSE

Before beginning to construct the smokehouse, dig two pits ten to fifteen feet apart. The first pit, upon which the smokehouse will sit, is used in the kippering process. It measures 2 by 2 feet and is 1 foot deep, with a 1- by 1- by 1-foot key at the back adjacent to the 2 by 2 foot square. This key, along with a door at the back wall, makes it easier to fill and clean the firebox. The second pit, which contains the fire for cold-smoking, measures 2 by 2 by 2 feet. If possible, the cold-smoke fire pit should be placed upwind from the smokehouse to aid in the draft of the stovepipe. Clear the ground around the first pit for the frame. The frame is made from 2 x 2 or 2 x 4s and measures three feet across the front and back and four feet on each side. Cut the front and back pieces the full 36 inch length while cutting the side 2 x 4s to 44½ inches. The 36 inch 2 x 4s then butt up to the ends of the 44½ inch lengths to give a full 48 inches. It will serve as the base for the four pieces of ¾-inch plywood that compose the walls.

A permanent smokehouse should also have a concrete floor around the pit. As the fish hang in the smokehouse, the oil that drips from them presents a sanitation problem. A concrete floor is easy to clean if you spread sawdust over it to soak up the oil, then simply sweep it away.

The next step is to build the sides. The fronts should be 72 inches high and the backs 68 inches. This 4-inch difference, or approximately one inch per foot, will allow rain to run off the roof. Use a full sheet of plywood so that you can make the walls 48 inches wide. Screw or nail a 4-foot 2 x 4 from the front to the back of each side 40 inches from the bottom to act as a brace and a surface upon which to lock the front door as shown in Fig. 24.

On the inside of each side, glue, then nail, a 2 x 2 from front to back along the top edge. The length of this 2 x 2 will be 4 feet, less ¾ inch. This board will help make an airtight joint at the roof and also provide a secure nailing surface for attaching the roof.

Next, as shown in Fig. 24, with glue, nail, or screw attach four 1 x 2 stringers to the inside of the wall to provide ledgers for

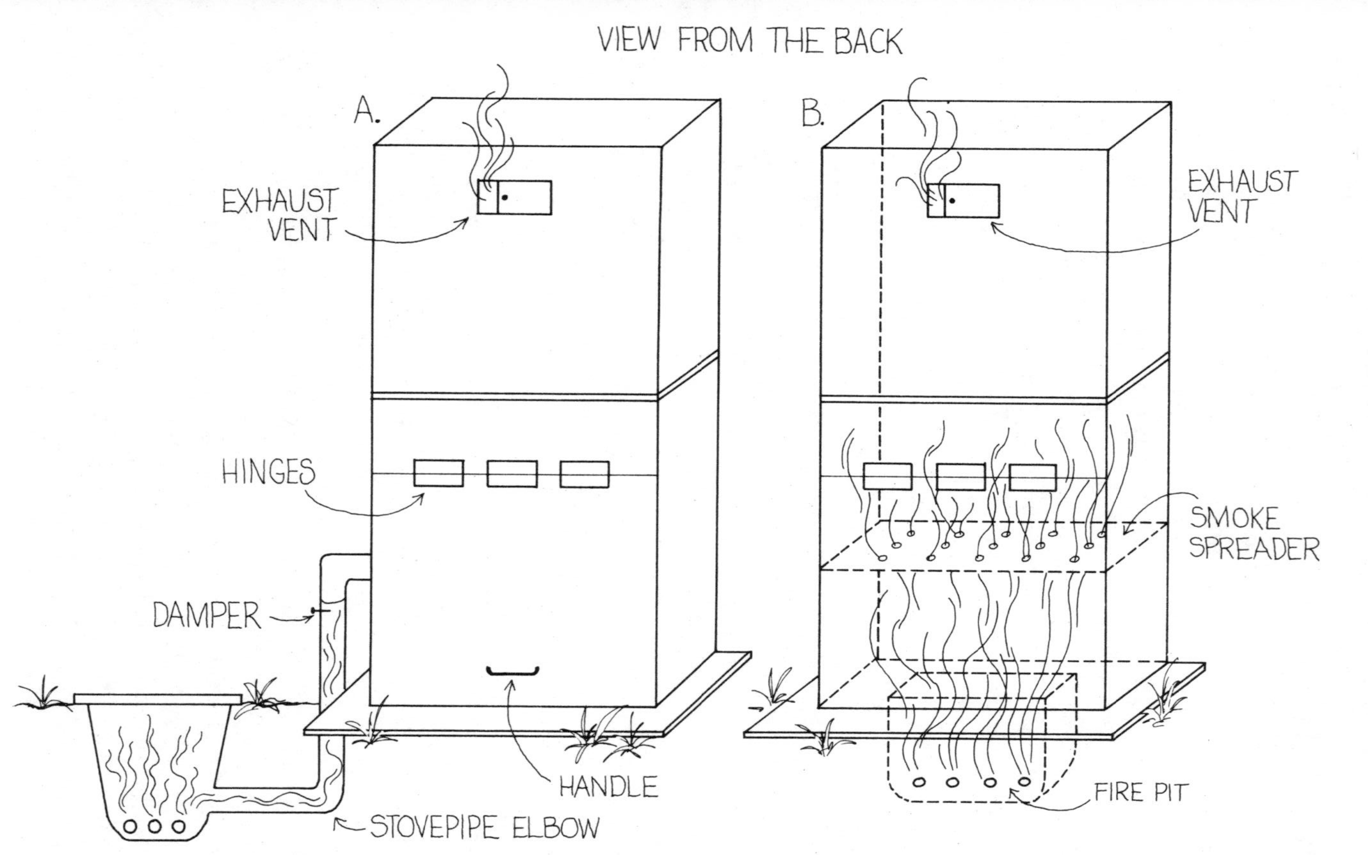

Fig. 23. This figure shows the two ways in which smoke is generated in the smokehouse. While only one smokehouse is actually built, the figure is divided into two views to illustrate the difference between hot smoking and cold smoking. The smokehouse in part A is used for cold smoking. The smoke is produced in a pit that is ten to fifteen feet from the smokehouse. The smoke travels through an underground stovepipe to its entry hole on the side of the house. The damper controls the volume of smoke travelling through the pipe.

Part B shows the smokehouse as it appears when it is being operated as a hot-smoking house. Build the fire in the pit directly beneath the smokehouse.

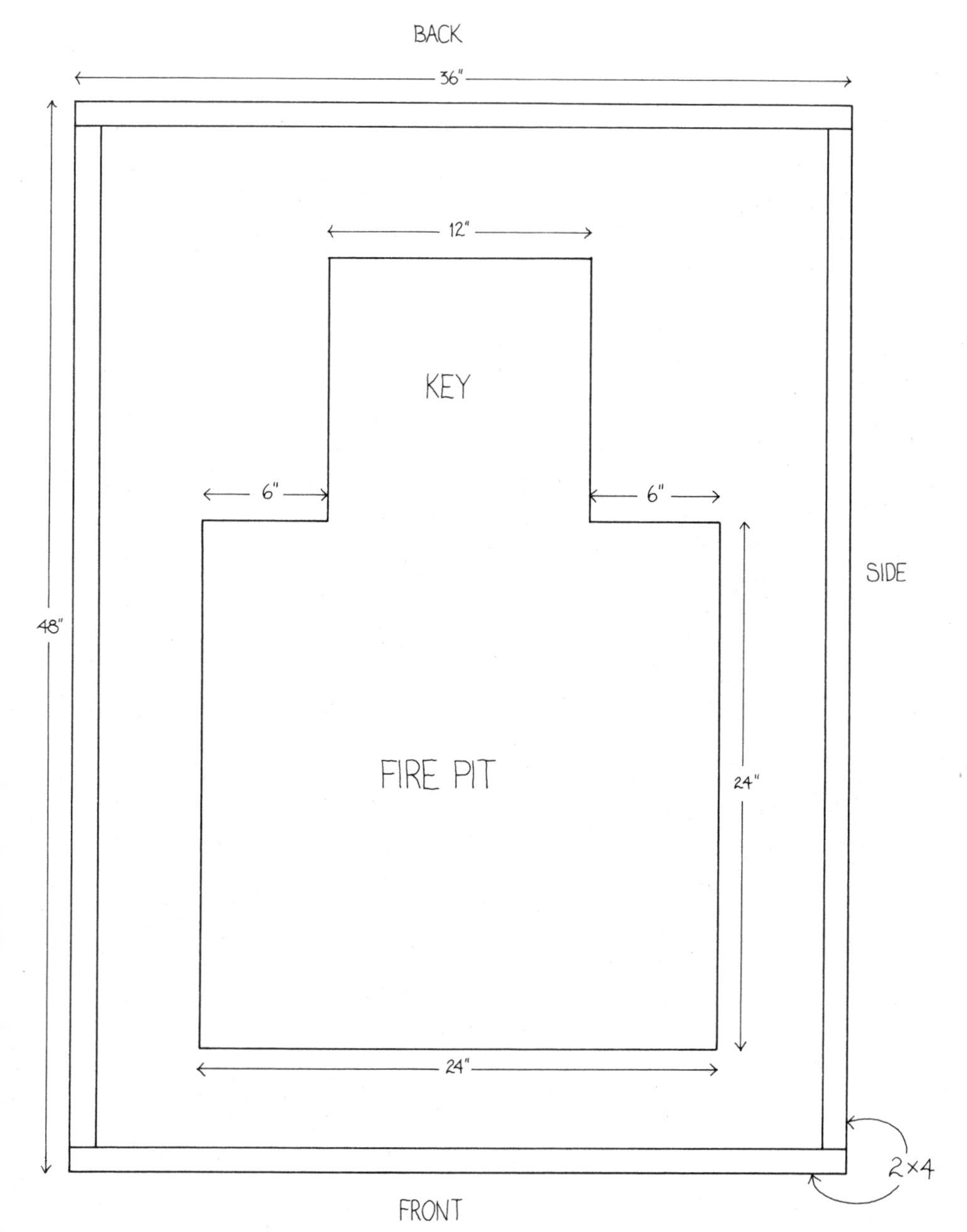

Figure 23C gives a detailed view (looking downward) of the shape of the firepit. Use the key to load the pit with wood and to clean out the ashes.

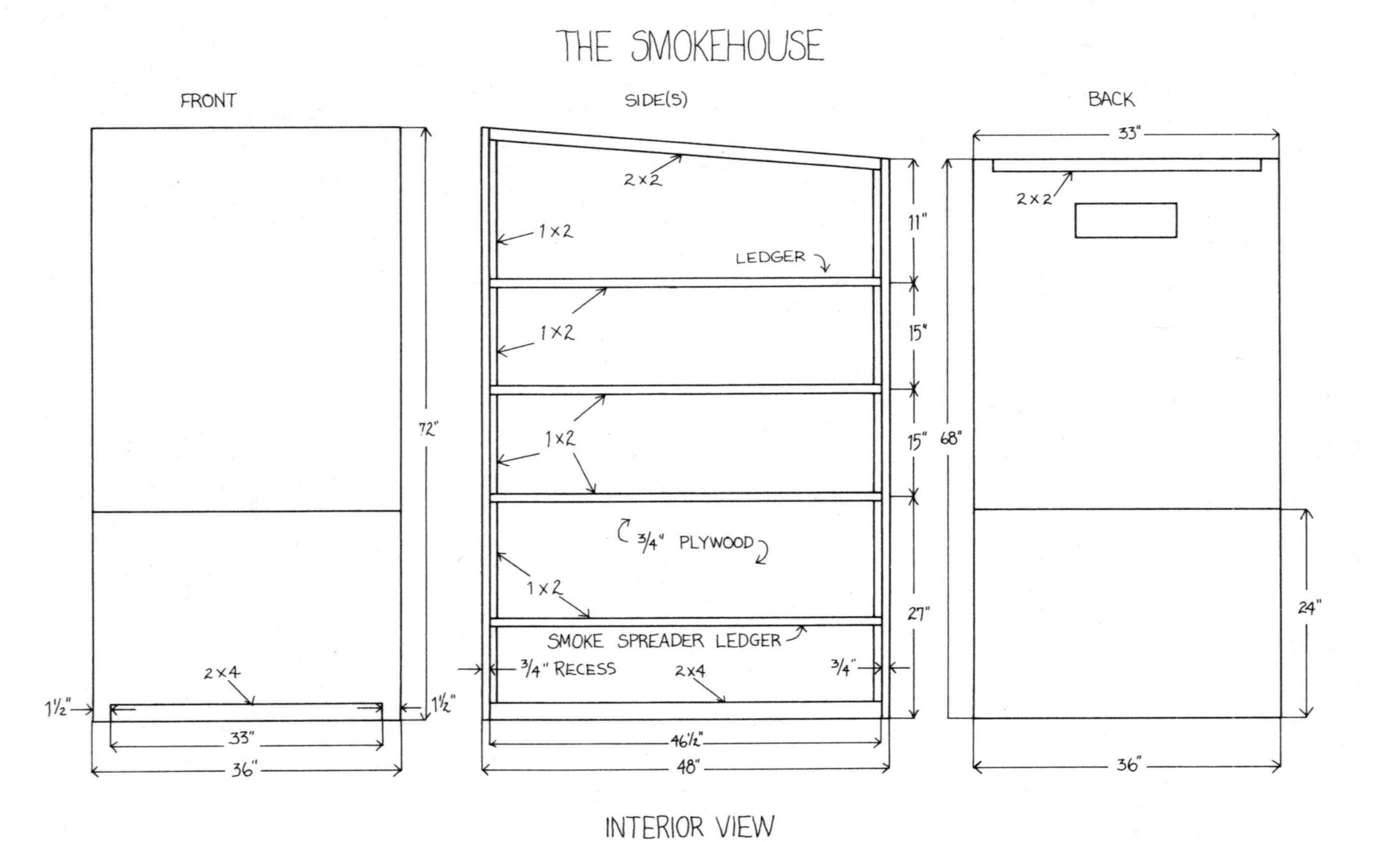

Fig. 24. The side of the smokehouse from the interior. The 1 x 2 strips run vertically between the ledgers.

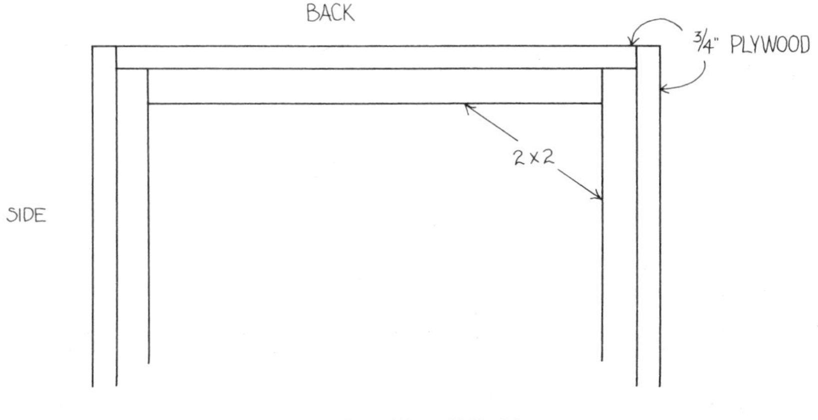

VIEW FROM THE TOP: THE BACK FITS BETWEEN THE SIDES

the fish racks or sticks for hanging fish. The lowest ledger, at 12 inches on center from the bottom, will hold the smoke spreader. The spreader, a removable 3- by 4-foot piece of ⅜-inch plywood with numerous one-inch holes drilled in it, diffuses the smoke evenly as it comes up from underneath. This ensures that all the fish receive a uniform amount of smoke.

The next three ledgers—at 27, 42, and 57 inches on center—are for the fish racks or hanging sticks. The space between each layer allows smoke to circulate freely between the fish. Spacing the ledgers minimizes the tendency to overload the smokehouse with too much wet fish.

When nailing the ledgers, allow a ¾-inch setback from the front and the back so that the door in front and the back piece will fit between the two sides. The ledgers will then, in effect, serve as doorstops. To help create a smoke-tight seal at the joints, attach 1 x 2 strips between the ledgers and from the ledgers to the 2 x 2s along the front and back of both sides.

The back of the smokehouse is 36 by 68 inches. Attach a 36-inch 2 x 2 to the inside of the back panel at the top edge, just like the 2 x 2s on the tops of the sides. Make a door at the bottom of the smokehouse by attaching one of the panels to the 36″ x 44″

back. Use three hinges and make sure the door swings out. Use hook-and-eye latches at each side to close the door. This door will be used for feeding the fire in hot-smoking and as an additional means of controlling the draft in cold-smoking. Glue and nail a 2 x 4 to the outside surface of the back 40 inches from the bottom to correspond to the 2 x 4s on the sides. Since the back will fit between the two sides, the 2 x 4 should be 41 inches long (2 by ¾ inch to compensate for the width of two pieces of plywood, plus 1¾ inches for each 2 x 4 on the sides). If you are making a portable smokehouse, the back should be screwed instead of nailed to the 1 x 2s on the inside of each side. Also screw the 2 x 4s on the sides to the 2 x 4 on the back.

The front piece of the smokehouse also has a door. This one, 48 inches high, is at the top of the panel. This larger door makes it easier to load and unload the fish.

Since the door is large in comparison to the ¾-inch plywood, the hinges should be reinforced. The simplest way to do this is to cut small plates from scrap plywood and, after lining up the hinges, drill through the plate and the door and bolt everything together. Make sure that the door still sits flush on the bottom piece of the front. You can make the bond stronger if you apply some glue between the plate and the door. To prevent the door from falling all the way to the ground attach a light chain from the upper corners of the door to the smokehouse.

As on the back, secure a 2 x 4, 41 inches long and 40 inches above the ground, to the front. Attach hook-and-eye latches to keep the door closed while the fish is being smoked. Put another latch at the top of the door to keep it tightly shut. The bottom two feet of plywood can be either nailed or screwed to the 1 x 2 strips and the sides.

Constructing the Roof

To construct the roof, nail and glue a 42- by 54-inch piece of plywood to the 2 x 2 frame at the top of the sides and back. Caulking the joints prevents moisture from entering and guarantees that

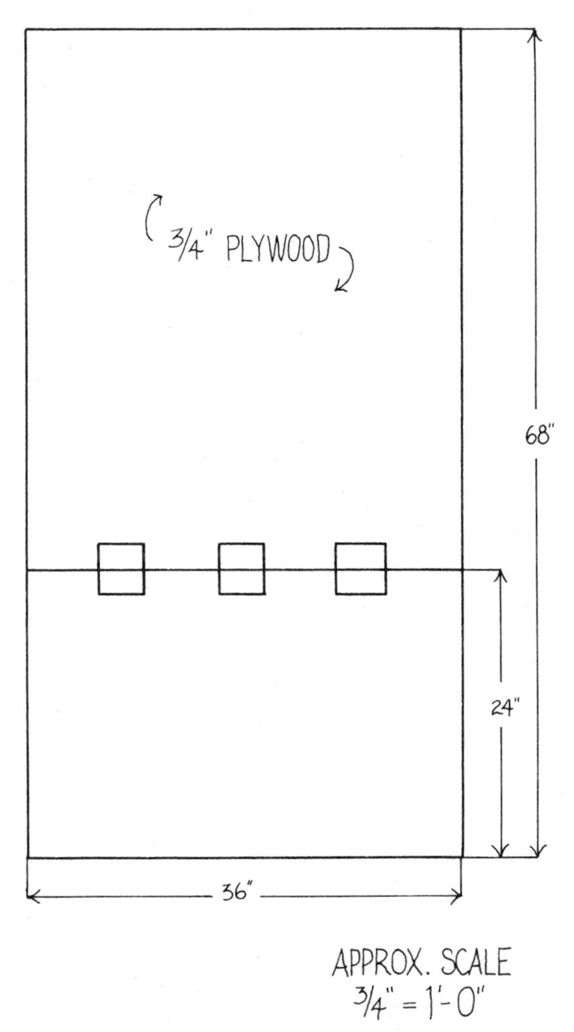

Fig. 25. A two-foot-high door on the back of the smokehouse allows easy access for cleaning. Attach three hinges to make a solid joint.

no smoke escapes except through an exhaust outlet cut at the top of the back. The 12- by 4-inch hole results in a stronger draft and better smoke circulation. Cover the outlet with a piece of canvas or make a sliding door for it so that you can adjust the size of the opening.

The outside of the smokehouse, and especially the roof, should be treated with wood preservative.

To make the smokehouse larger, merely lengthen the dimensions from front to back, or from side to side.

Installing the Stovepipe

On the side of the smokehouse closest to the cold-smoking fire pit, cut a hole 6 inches in diameter. The lower edge should be 6 inches above the ground. A series of elbows and straight sections of 6-inch diameter stovepipe connect the smokehouse to the fire pit through an underground trench. The length of the trench determines how many sections of stovepipe you'll need. Elbows make right-angle bends. You need at least two, more depending on how the pipe is laid. The damper should be in the section of pipe closest to the smokehouse.

Ventilating the Smokehouse

Air circulation is the key to smoking fish. If the smokehouse doesn't draw enough air by itself, a system of fans must be devised to maintain proper air flow, both for drying and smoking the fish. A blower at the opening of the stovepipe from the cold smokepit will circulate air from the bottom of the smokehouse to the top.

Since the air-smoke mixture comes in from the side of the smokehouse, a sheet-metal baffle should be placed at the opposite side to deflect the air upward. This arrangement will produce a circular air flow. Hang pieces of ribbon or newspaper strips in the smoke chamber to show the degree of air movement. To achieve a balanced air flow, either insert more baffles or bend the sheet metal to a different angle. In cold-smoking with a blower, the smoke spreader should be removed.

BUILDING THE FIRE

The fire in the cold-smoking pit can be built in many ways. A simple fire can be made from hardwood logs about four inches in diameter. Almost any type of wood will work except resinous woods, such as fir, pine, cedar, and the other conifers. In the Northwest, alder is the most commonly used wood, but hickory, maple, walnut, oak, and ash also produce a good smoke.

Another way to produce the smoke is by burning sawdust. Light a pile of hardwood sawdust or chips with a piece of news-

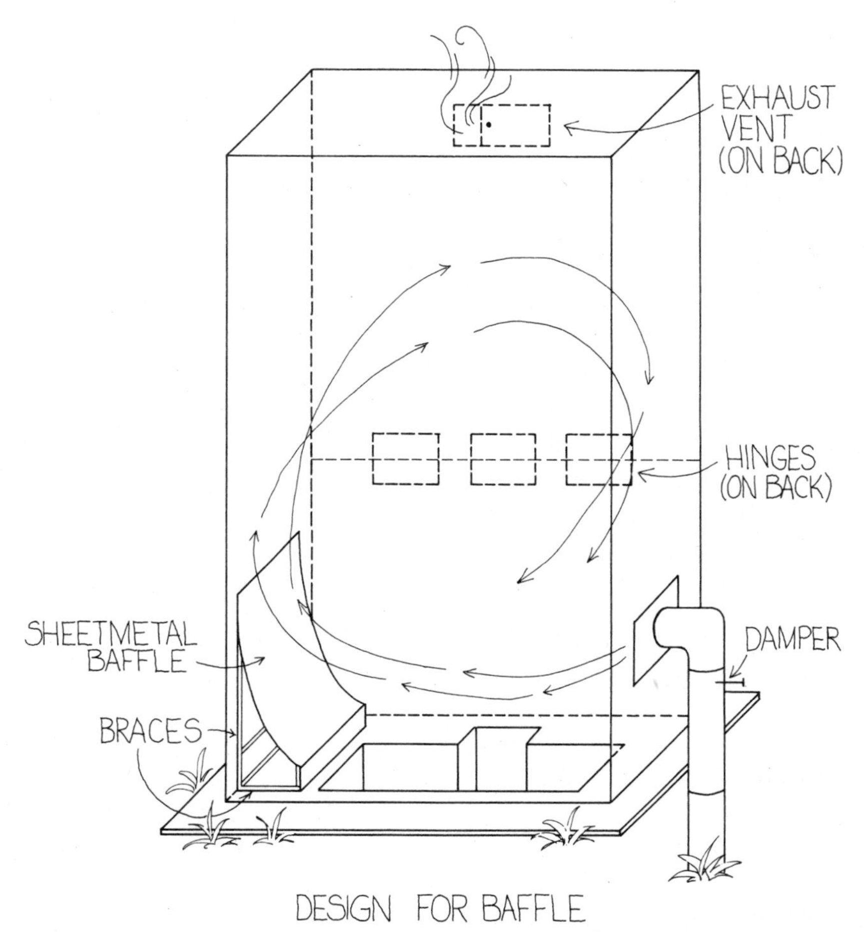

Fig. 26. The baffle along the left inside wall of the smokehouse causes the air to move upward after it enters through the stovepipe. A fan or blower at the opening of the stovepipe increases the air velocity.

paper and cover the fire pit with a metal garbage can lid. If you use this method exclusively, build the fire box in the shape of an inverted cone. The side closest to the stovepipe should be open so that the sawdust will settle to the bottom and burn evenly.

Instead of building a fire, you might want to place an electric hot plate in the fire pit with hardwood chips in a metal pan on top of the burner. However, sometimes the incomplete combustion that results from this method produces aliphatic acids, phenols, and tars, which impart an acrid taste to the fish. Gray rather than blue smoke indicates incomplete combustion. Because the hot plate method is easy to control, it is worth trying.

The fire in the smokehouse fire pit can be built in the same ways as the cold-smoke fire. Another method is to lay a bed of charcoal briquets in the pit and set 4-inch logs on top of them to smolder. Use the back door to regulate the temperature. Whatever method you choose, you must tend the fire at regular intervals. It's not possible, of course, to build a fire that will hold a constant temperature for six or eight hours.

The smoke spreader should be in place when the smokehouse fire pit is used. To avoid burning the wood too fast or producing a flame high enough to burn the spreader (and the rest of the smokehouse), the blower should not be used, although it can be used to dry the fish before building the fire. An overhead fan, secured at the roof, can be used to circulate the air. Rotate the racks of fish periodically during the smoking process to produce uniform results.

MAINTAINING TEMPERATURE

In parts of the country where high temperatures and high humidity are a problem in the summer, it is best to do most of the smoking (particularly cold-smoking) in the fall and winter months. Cold air carries much less moisture than warm air, so condensation in the smokehouse is minimized. The drier air will also dry the fish more quickly.

Air that has a relative humidity of 100 percent at 40° has a 50-percent relative humidity at 60° and a 25-percent relative humidity at 80°, assuming that the amount of moisture remains constant. In other words, 80° air holds four times more moisture than 40° air.

Another factor to consider is that it is much easier to heat the smokehouse to the proper temperature than it is to cool it down. Fish that are to remain in the smokehouse for a week will be fine at 70° but will spoil if the temperature reaches 90°. To reduce the effects of temperature differences between the inside and outside of the smokehouse, the walls and ceiling should be insulated.

To maintain a 70° temperature—and an even higher temperature for kippering—you may have to put a thermostatically controlled heating element or small space heater in the smokehouse. An electric hot plate placed in the pit under the smokehouse does an adequate job of bringing the air in a small smokehouse up to temperature for cold-smoking. In hot-smoking, the fire itself heats the air. With the 2- by 3-foot door at the back of the smokehouse, regulate the amount of cool air that is allowed in.

CLEANING THE SMOKEHOUSE

After the smokehouse has been used for a while, the interior, the racks, and the inside of the stovepipe will begin to accumulate a thick, sticky coat of creosote and other tars from the smoke. It's not a problem on the wood, but it must be cleaned from the motor and other electrical parts of the fan and heater. Use hot water and strong detergent to clean the motor. Be careful not to get water inside the motor. If not cleaned the motor may bind and burn out, or the heater may ignite and burn the smokehouse. If the creosote deposits in the stovepipe ignite, the roaring fire may burn through, also destroying the smokehouse.

The sticks and racks used to hold the fish should be washed with soapy water as part of a regular routine.

4

Canning

Canning, or sealing food inside airtight containers and processing them with heat, was developed by the Frenchman Nicholas Appert between 1795 and 1810. He was awarded a prize of twelve thousand francs by the French government for his discovery. But Appert didn't understand the principle that made his process work, and it wasn't until Louis Pasteur did his research that the process became clear. The canning procedure is still sometimes referred to as Appertization.

Canning fish at home, like any home canning, is simple, safe and economical if a few basic guidelines are followed—and followed *without* deviation and without taking *any* shortcuts. The risks of carelessness and unsanitary conditions can lead to illness and even death. Botulism, from *Clostridium botulinum,* is a real danger in improperly canned fish. It is particularly insidious because it is so difficult to detect in the canned product. And it is usually fatal.

Use only wholesome fresh or frozen fish in canning. The canning process will not improve the quality of poor or mediocre fish. And spoiled fish has a high bacteria count that the pressure canner will not completely destroy. The fish themselves should be free of slime and dirt. In a freshly killed fish the meat is sterile. If the outside of the fish is contaminated, germs will be introduced in the tissue when the fish are cut.

Two conditions are responsible for most of the spoilage in canned fish. The first occurs when the fish are not cooled quickly enough after they are caught. The second occurs when the fish are packed in jars and waiting to be processed in the pressure canner. Work quickly to keep this time interval short.

CANNING FROZEN FISH

Frozen fish make a good canned product if they have not been frozen for too long a period of time. If you want to can frozen fish that was brined first, do not salt it before canning. Improperly frozen fish turn yellow from oxidation and will have a "freezer" odor when the jar is opened.

Before freezing lean fish that are to be canned, brine them for fifteen seconds in one gallon of water in which one and one third cups of salt have been dissolved. Fish with a high oil content will oxidize faster if they are brined, so they should be frozen without brining. For oily fish, use one of the brineless techniques in chapter 1.

Thaw the frozen fish and soak them for ten minutes in a 70° brine solution or in the solution specified in the recipe you are using. The brining will help to reduce some of the toughness in the fish and minimize the formation of curd. Curd is water-soluble protein that coagulates into white chunks on the surface of canned fish.

PREPARING JARS, LIDS, AND SCREW BANDS

Use only canning jars that are manufactured specifically for home canning, such as Ball Mason jars. Make certain that they are clean and free of chips and cracks, especially around the sealing edge of the rim. Even scratches on the glass can cause the jars to break when they are subjected to heat. Commercial food jars, such as those for peanut butter or mayonnaise, should never be used for home processing. The glass is not thick or tempered enough to withstand the pressure-cooking temperatures.

The recipes in this chapter are only for *pint* jars; larger jars require different cooking times and pressures to reach the temperatures needed to kill disease-causing bacteria. The size and shape of the jar as well as the distance to its center are important factors in determining the processing temperature. These factors also determine the length of processing time. The processing times and temperatures given for pint jars are not interchangeable with any other size jar.

The metal canning lids and screw bands should be appropriate for the jar. Read the manufacturer's instructions for your particular jars to determine the proper lids and how they are to be sealed. The step in the canning procedure where they are tightened varies with different types of jars.

Several kinds of lids and screw bands are available for home-canning use. Since each type is different, be sure you know whether it is possible to reuse them. The most commonly used lids have an inner ring of sealing compound. Since the sealing compound can be damaged easily, never reuse a lid that has sealed on a previously canned jar. The metal screw bands can be reused if they are rust-free and not dented or bent.

Cleanliness is very important in all stages of the canning process. It is important that counter tops, cutting boards, stove, knives, and all utensils be absolutely clean. Floors and walls should also be clean and free of grease. Also make sure that your hands and clothes are clean.

Wash the jars and lids in hot soapy water and thoroughly rinse them in hot water. Put the jars in a pot and cover them with water. Boil the jars for several minutes. Remove the kettle from the heat and place the lids in it. Do not boil the lids or you may damage the sealing compound. Leave the jars and lids in the hot water until each one is ready to be used. In this manner, recontamination is avoided. Keeping the jars hot also prevents breakage when they are placed in the hot canner.

PREVENTING THE GROWTH OF BACTERIA

Contamination in canning comes mainly from bacteria and their toxins, but molds and yeasts also contribute to product spoilage. Heat is the most efficient bacteria killer, but even at high temperatures some bacteria will remain active. Three main types of bacteria are of concern to the home canner. *Salmonella* is the rod-shaped bacteria responsible for food poisoning. Although these bacteria are inactive below 45°, they are not killed until the temperature reaches 140°. A more dangerous bacteria is the spherically shaped *Staphylococcus aureus*, which produces toxins in the 40° to 140° temperature range. Several hours at 212° (the boiling point of water), or thirty minutes at 240° are required to kill this bacteria. The third bacteria, and potentially the most deadly, is *Clostridium botulinum,* an anaerobic, rod-shaped bacteria found in soil and on the ocean floor. When these organisms are threatened, they form spores that are capable of suffering great variations in humidity, temperature, and salinity.

While *botulinum* are the most toxic bacteria, they are not the most heat resistant. *Clostridium sporogenes* and *Clostridium putrefaciens* are even more resistant to heat than *botulinum,* but they are not as deadly. The only way to keep them from growing in the canned food is not to introduce them in the first place.

Simple washing removes most of the bacteria from food, but the spores are destroyed only at high temperatures. Oil in the pack surrounds the spores with a protective coat and makes them less vulnerable to heat. However, the oil does cause the spores to lie dormant in the processed jar. The temperature and length of cooking time required to kill these spores vary from product to product, depending on the acidity and density of the food, so no single temperature is given to eliminate them. As a general rule, the number of bacteria destroyed is increased by a factor of ten for every increase of 18° Fahrenheit. The container size, temperature (usually 240°), and the processing time are given in each recipe.

The processing time in home canning is a balance between the minimum necessary to kill the bacteria and the maximum allowed before the food breaks down.

Specific recipes are given for different types of fish because of the variation in fish characteristics. If a certain species of fish isn't listed, the fish probably doesn't can well. Ling cod, for example, doesn't respond well to canning because of what is called the Maillard reaction: The fish turns brown and assumes an off taste, and the protein itself becomes insoluble. The problem is compounded during canning because high temperatures increase the rate of the reaction. Fish that are high in iron, such as abalone, sometimes turn blue or black when canned. The discoloration is a result of a sulfide reaction. Since sulfides are present in fish tissue, it is a difficult problem to overcome. Water used in the canning process that is high in iron makes the problem even more acute. If you are in doubt about a particular type of fish, test-can a small batch before canning the whole catch.

Molds, which produce mycotoxins, and yeasts are both inactive below 32°, and they die at 190°.

THE IMPORTANCE OF THE pH BALANCE

The pH of foods (rated on a scale of 0–14) measures the amount of acid in foods. A pH of 7 is neutral, while anything with a lower pH is acidic. A pH between 4.6 and 7.0 is considered to be low acid. Salmon has a pH of 6.1–6.3; tuna has a pH of 5.9–6.1. Both are classified as low acid. [Any pH number above 7 indicates the substance is a base.]

In canning low-acid foods, a pressure canner is essential because it achieves the high temperature necessary for sterilization. Water-bath canners, which can be used for tomatoes, pickles, and jams and jellies, should never be used for low-acid foods.

At sea level in an open pot, water boils at 212° and will stay at that temperature so long as there is water in the pot. Only after a lid is clamped on the pot will the vapor pressure inside bring the temperature above 212°. Most of the recipes in this book will use a pressure of ten pounds per square inch to raise the temperature to 240° in the canner. Fifteen pounds of pressure at sea level produces 250°.

Ten pounds of pressure is necessary to achieve 240° at sea level. At 2,000 feet, eleven and a half pounds must be used; at 3,000 feet, twelve pounds are required; at 4,000 feet, twelve and a half pounds are required to achieve 240°; and at 5,000 feet thirteen pounds are required. In other words, for every thousand-foot gain in elevation (after 2,000 feet), an additional half-pound of pressure is needed to maintain the same temperature.

PRECOOKING BEFORE CANNING

Home-canned products are extremely beneficial to people on salt-free, low-fat diets because both salt and fat can be easily controlled. Salt performs no preservative role in canning and serves merely to enhance the flavor of the fish. Never use iodized salt: It can result in bitterness, and it also clouds the canning liquid. Fish with a high oil content can be precooked to remove some of the oil from the meat. The addition of vegetable oil is optional for these fish.

When fish are cooked before they are canned, several changes occur. The flesh loses water and gelatin, the substance in fish skin and tissue that is liquid when hot, but a sticky solid when cooled. For example, tuna is approximately 68 to 76 percent water before cooking, whereas afterward it is about 60 percent.

The protein in the tissue shrinks, making it easier to remove the skin from the fish. Pull off the skin before it has cooled since the skin sticks tightly to the meat after it cools. Precooking also helps to remove the sulfides, which cause discoloration in the canned fish.

It is not essential to bone most fish before canning them because processing makes the bones soft enough to eat, and the bones provide an added source of calcium.

OPERATING THE PRESSURE CANNER

There are two basic types of pressure canners in use today. The first has a dial gauge to indicate the pressure (and thus the temperature) inside the canner, while the second has a weight gauge that allows just enough steam to escape to maintain the desired pressure. The weights come in five-pound increments—five, ten, and fifteen being the most common. Above 2,000 feet in altitude, the fifteen-pound weight must be used when ten pounds is called for at sea level to achieve 240°.

No matter what type of canner you use, always read the manufacturer's operating instructions first. Practice with the canner before you start to can so that you will have an idea of the required stove setting for maintaining the proper pressure. If the pressure ever drops below the one specified, the timing must be started over again and the fish recooked for the entire period. To process fish properly, the temperature must be continuously maintained above the level called for in the recipe. There can be no deviation from this rule, so be sure you know how your particular canner works.

For the safe operation of the pressure canner, it must be kept clean. Not only should the rims of the kettle and lid be wiped clean of particles that might interfere with the seal, but the gauge should be cleaned to ensure that it operates properly. A piece of wire, a toothpick, a short string, or a pipe cleaner can be used to clean the vents in the lid.

The dial gauge should not be immersed in water but should be wiped clean with a damp cloth. Before the canning season begins each year, have the gauge tested for accuracy. Consult your local county extension agent for help in testing the dial gauge.

Place a metal rack in the bottom of the canner and fill the canner with two to three inches of water. Load the canner with the pints of fish. If the canner is tall enough, put another rack under the second tier. The racks guarantee good steam circulation so that all jars cook equally. The bottom rack prevents the jars from coming in direct contact with the heat. (See Fig. 27.)

Venting the Canner

To vent the canner, close the lid and turn up the heat. Bring the water to a boil, then reduce it to the level necessary to maintain the proper steam pressure. When the vapor begins to escape

Fig. 27. When placing two layers of jars in the pressure canner without a rack between them, stagger the jars to ensure proper steam circulation. Never place one jar directly on top of another.

Fig. 28. This dial gauge reads ten pounds of pressure. It is crucial that you maintain a steady pressure during processing.

from the vent, note the time. Allow the steam to escape for ten minutes. If the canner isn't vented, the air inside that is at one atmosphere of pressure will not mix properly with the steam and will inhibit the buildup of the steam pressure that causes the temperature to rise.

After the ten minutes have passed, close the vent to bring up the steam pressure. It takes five to ten minutes for the pressure inside the canner to come up to ten pounds. If you have a dial gauge, it will indicate the pressure at any given time. Write down the time when the proper pressure is achieved and keep the pressure as steady as possible at that level. If a recipe calls for ten pounds of pressure, the dial must *never* go below ten. (See Fig. 28.)

A word of warning!—If the pressure goes down, the temperature goes down, so the time must be noted and the count started again when the proper pressure is reached. The time specified in the recipe is counted *from the time pressure is reached* (ten pounds in this example) until the cooker is removed from the stove and the pressure allowed to drop.

If you raise the pressure too high, the water in the jars will begin to seep. High temperatures also cause the nutrients in the fish to break down at a faster rate.

The weight gauge is similar in function to the dial gauge. Displace the air in the closed pressure canner by letting the steam escape through the open vent for ten minutes. Then place the proper weight (five, ten, or fifteen pounds) on the vent. As the pressure in the canner rises, the steam will cause the weight to jiggle and rattle. The proper rate of jiggling is two or three times per minute. When this rate is reached, write down the time and begin the count specified in the recipe. Don't let the jiggling become too slow or the count *must* be started over. Too rapid jiggling results in too much water being converted to steam. In this case, the canner could go dry before the time is up.

After the specified time from full pressure has elapsed and the pressure cooker has been removed from the stove, wait at least thirty minutes before opening the lid. The pressure must drop sufficiently first. A dial gauge will indicate when zero pounds of pressure has been reached, but a weight gauge will not. Lift the weight slightly (but not with your bare fingers) and watch for the steam to escape. As it escapes, it will make a hissing noise—either from air rushing in or steam rushing out. A noise alone will not indicate that the pressure has dropped to a safe level. A rapid pressure change in the canner can result in fluid seepage from the jars, so allow the cooker to lose steam pressure at its own rate.

When the pressure reaches zero pounds, wait a few more minutes before opening the lid. It is always a good idea to open the far side of the lid to let any steam that remains escape *away* from you. A steam burn can be extremely painful. If the liquid in

Fig. 29. Use the canning jar to measure the length of the fish. Don't cut the fish to the full length of the jar; allow head space.

the jars is still boiling, wait a few minutes before removing them with a jar lifter.

FILLING THE JARS

Wide-mouth jars with a small shoulder are convenient to use because they are easy to pack and permit easy removal of the food. To fill the jars, mark a piece of cardboard or a measuring stick equal to the depth of the jar, or use the jar itself as a measure and cut off the first piece. (See Fig. 29.) For salmon and other round fish, place the fish on the cutting surface with its back facing up. Cut through the backbone between two vertebrae, roll the fish onto its side, and cut through the belly wall. Pack the fish in the jar with the skin side out. The fish should be packed tightly without being crammed in. (See Fig. 30.)

Run a sterilized knife or a spoon around the inside of the jar to remove any air bubbles. Air bubbles will not transfer heat at the same rate as the liquids in the jar, and as a result underprocessing may occur. Use broken pieces of fish to fill empty spots in the jar. Wipe the rim with a clean, damp cloth to remove any oil that would prevent the lids from sealing, and make sure that no fish or bones hang over the edge of the jar. The rim of the jar should be completely clean.

The Question of Head Space

There are two schools of thought concerning the amount of head space to allow in the jar. One says that no head space is required with fish (the jar can be filled to the rim), since they will shrink enough during processing. If the fish is cut before cooking

Fig. 30. Pack the fish in sterilized jars. The skin should be on the outside, with the bone to the inside. Remove air pockets by running a knife around the inside of the jar.

to allow head space, too much space may result after the shrinkage. The other school says this phenomenon isn't precise enough to rely on: The fish can be cut to jar length and soaked in a brine of one cup salt to one gallon of water for an hour. The fish will firm up and shrink so they can be packed with one inch of head space.

Brining has two other beneficial effects. The amount of curd that forms on the top of the fish is reduced because the tissue holds more water (and thus the water-soluble proteins that compose curd) after brining. And the meat of the brined fish doesn't stick to the inside of the jar. Recipes are given for canning both brined and unbrined fish in the following section.

Head space is important for two reasons. When the jar is packed, it will have either a small amount of air at the top, or it will have fish and a small amount of fluid up near the rim. As it is heated, the air expands and escapes from the jar, forming a vacuum. If the head space is too great, not enough air will escape from the jar. Either the lid will not seal, or worse, the lid will seal with air still inside. Any aerobic bacteria present would then grow and multiply.

Head space is also necessary to prevent spillage when the liquids in the jars boil. If the liquid spills out, particles of fish or oil will remain on the rim of the jar, preventing a good seal and permitting air into the jar. Bacteria and other contaminants will enter with the air.

After the jars are packed and cooked, remove them from the pressure canner with a jar lifter and place them in an open area to cool. It is easy to break hot jars and particular care must be taken to avoid cracking them. Set the hot jars on dry towels or on a half-inch stack of newspaper in a draft-free area. Leave a few inches of space between jars for good air circulation while cooling. It is important to cool the jars as quickly as possible since the fish continue to cook in the hot liquid, but precautions must be taken to avoid breaking the glass. Never attempt to cool the jars in cold water.

TESTING FOR SEALS

As the jars cool, they will make a loud popping sound when the lids seal. However, *listening* for a seal is not a practical test.

After the jars have reached room temperature (about twelve hours), test the seals. A jar that has sealed properly will have a depression in the center of the lid. Sometimes this depression will be slight, but if the jar is sealed the lid will stay down when it is pressed. A spoon tapped on the lid of a sealed jar will produce a clear ringing sound, while one that is tapped on an unsealed jar will make a dull thud.

The most reliable test is whether or not the jar can be picked up by the edge of the lid (not the band). Be careful not to raise the jar more than an inch or so; if the seal is not tight, the jar will drop.

Most jars of fish will form a depression in the center of the lid as soon as they are removed from the pressure canner. Any jars that do not form the depression should be separated from the rest of the pack and checked periodically. After three hours from the time of removal from the canner, the jars should be sealed. If they are not, you must assume that they are not going to seal. At this point, two options are available: to cool the jars to room temperature and put them in the refrigerator, or to reprocess them. The jars that go to the refrigerator must be treated like any other cooked fish. They will keep for several days in the refrigerator, but will be highly perishable at room temperature.

Jars of unsealed fish can not be reprocessed if they have been out of the pressure canner for more than three hours. Begin reprocessing by determining why the jar did not seal the first time. Check the rim of the jar for cracks or chips in the glass. Make sure there is adequate head space and that pieces of fish aren't hanging over the edge of the jar. Wipe the rim to remove any oil that may prevent a seal. Put on a new lid. Lids may not be re-used. Mark the jar to show that it has already been cooked one time.

Cook the jars of fish in the same manner as done the first time. Bring the water in the canner to a boil and vent the air from the canner. Adjust the heat and let the pressure rise in the pressure cooker to the proper level. Cook the fish for the full amount of time specified in the recipe.

Reprocessed fish will be lower in nutritional value and flavor and will have a softer texture than fish canned only once.

After processing, if the liquid in the jars doesn't cover all the fish, don't open them. The exposed fish may oxidize faster than the rest of the fish, but they will not decompose. There are several reasons why liquids leave the jars during the cooking phase. Jars that are packed too tightly or with insufficient head space have a tendency to boil over, as do jars with improperly applied lids. Sudden pressure fluctuations (because of temperature changes) in the pressure canner also cause liquids to escape.

When the jars have cooled and sealed, wipe them clean and label them with the type of fish, how it was packed (whether it was brined or precooked, and whether salt, water, or oil was added), the length of the cooking time and the pressure, and the date packed. Number the jars as parts of a series, such as 1/10, 2/10, 3/10. The numerator is the jar number and denominator is the total number of jars in the batch. This numbering system makes it possible to find all the jars of a particular batch if a problem is later discovered with the fish in that batch.

STORING THE JARS

Store jars of canned fish in a cool, dry, dark place. Don't can more than one year's supply of food at a time, since nutritional loss is too great after that period. Any temperature above 65° causes an increase in enzymatic activity in the fish tissue and results in deterioration. Light also increases the rate of reaction, causing the fat in the fish to become rancid. In addition, light brings about color changes in the flesh, creating a paler product. Vitamins, especially riboflavin (vitamin B_2), lose their nutritive value when exposed to light.

The jars must be dry to prevent rusting on the lids. Rapid

changes in temperature that occur during storage can cause condensation on the jars which will allow the lids to rust and the seal to break. If the seal becomes broken during storage, the food should be disposed of immediately where no person or animal can eat it. DO NOT taste the fish to see if it is good.

Check the jars for spoilage, and make sure the seal is firm before opening the jar. There should be no bubbles around the edge of the jar. Neither should there be any mold on the seal or on the food. There should be no little bubbles floating in the jar or at the top of the liquid. The liquid itself should be clear and not cloudy. The lid should maintain its depression in the center and have no bulges. Any negative sign in these inspections warrants disposal of the fish.

When the jar is opened, no liquid should come spurting out. A yeasty smell or a musty odor indicates decay. Slimy, soft, moldy, or shriveled food should be destroyed. If in doubt, throw the food away and forget it. There is no reason to risk your family's health over a pint of fish.

A final measure to denature any bacterial toxins is to boil the food for fifteen to twenty minutes, or to bake the fish in a 350° oven until a thermometer registers an internal temperature of 185° (about thirty minutes). Allow the fish to sit at room temperature for thirty minutes so the heat can disperse throughout the fish.

CANNING RECIPES

Salmon and Shad

Cut the fins and tail from the cleaned, headless fish. Without filleting or boning the fish, cut across the backbone so that the pieces are equal to the length of a pint jar.

Prepare a brine with eight ounces of salt to one gallon of water (approximately 22° on the salinometer) and soak the fish in the brine for one hour. A gallon of brine is sufficient for soaking twenty pounds of fish. Do not reuse the brine: The salt draws fluids from the fish and thus reduces the concentration of the fluid. Drain the liquid from the fish as it is packed.

Place the fish in a firm pack with the bone to the inside and the skin to the outside. Do not add water to brined fish. Close the jars and process the pints for one hour and fifty minutes at ten pounds pressure (240°). As a general rule, twenty-five pounds of fish (live weight) will yield approximately twelve pint jars.

Another method is to prepare the fish as described above and pack them directly into the jars without brining them. Pack the fish firmly in the jars until they are just below the level of the rim. Add a teaspoon of salt to each jar and fill it with boiling water. The boiling water helps reduce the number of microorganisms on the surface and exhaust the air from the jar. After closing the jars, process them for one hour and fifty minutes at ten pounds of pressure.

Fish can also be canned at fifteen pounds of pressure. At altitudes less than 3,000 feet, process the jars in a 12-quart, or larger, pressure canner for eighty minutes; above 3,000 feet, process them for one hour and fifty-five minutes. At altitudes above 7,000 feet, the jars must remain in the pressure canner for two hours and thirty-five minutes.

Tuna, Albacore, and Mackerel

Tuna, albacore, and mackerel contain strongly flavored oils that are removed by precooking. Dress and wash the fish and trim away the thin belly strips.

Put the fish in the pressure canner in two layers to avoid making too solid a mass of fish. Use a rack between layers to permit the steam to circulate.

Cook the fish for two hours at ten pounds pressure, or place them in a roasting pan and cook for one hour in a 350° oven. Place a pan of water in the oven for moisture. This will keep the fish from becoming too dry. Cook the fish until the meat above the backbone—the thickest part of the fish—is 155° to 165°. Don't overcook the meat.

Cool the fish thoroughly and as quickly as possible to prevent bacterial growth. You can cool the fish in the refrigerator until it is room temperature, or you can run cold water over it. As soon as the fish are cool enough to touch, scrape off the skin and remove the backbones. The skin will be harder to remove once it is completely cold.

After removing the skin, continue to cool the meat to make it firm. Use only the white meat of the fish and discard the dark meat.

Cut the fish so there will be half an inch of head space and pack the flesh firmly in pint jars, adding one-half a teaspoon of salt to each jar. The fish must be cut shorter than the length of the jar since precooked fish will not shrink as much during processing as raw fish. If desired, add three or four tablespoons of heated vegetable oil, or an equivalent amount of boiling water.

Close the jars and process them at ten pounds for two hours. An alternative is to process them at fifteen pounds pressure for the same amount of time as listed for processing salmon at fifteen pounds.

Mackerel, Lake Trout, and Whitefish

Prepare the fish in the same way as salmon. After the brined fish are drained, pack them head to tail (to utilize fully the jar capacity and to avoid air spaces) in pint jars, leaving a quarter inch of head space.

Place the still-open jars in a kettle and cover them with a brine solution of four ounces of salt to one gallon of water. Boil the jars of fish uncovered for fifteen minutes, then drain them, inverted, over a wire rack for several minutes.

Close the jars and process them at ten pounds pressure (240°) for one hour and forty minutes.

Clams

Keep the clams alive in a ten-percent brine for twenty-four hours. You do this by cutting a saturated brine (three pounds salt to one gallon water) with nine parts of water to each part of brine. Remove them from the brine and scrub the clam shells and place those that are tightly closed in a steamer. Discard any clams whose shells have opened; these clams are dead and may be spoiled. Steam the clams until the shells open.

Open the steamed clams and reserve the liquor. Heat the liquor in a saucepan and reduce it until it is two-thirds the original volume. Snip off the siphons of the clams and throw away the dark portion of the intestinal system.

Wash the clam meats in a weak brine made by mixing one-quarter cup of salt per gallon of water. Mix a solution of either two tablespoons vinegar per gallon of water, three teaspoons lemon juice per gallon of water, or half a teaspoon of citric acid per gallon of water and bring it to a boil. Blanch the clam meats in the boiling solution for one to two minutes.

Pack one and three-quarters cups of clams, either whole or minced, into each pint jar and fill them to within a half-inch of the top with the boiling clam nectar. Put the lids on the jars and process them at ten pounds pressure (240°) for seventy minutes.

Smoked Mackerel

Mackerel may be given a light smoke before preparing them for canning. The smoke taste intensifies during the canning process, so oversmoking must be avoided. Cut the smoked fish into strips one to two inches wide and the length of the pint jar.

Pack the strips tightly to make a solid pack and cook them, uncovered, in the pressure canner for twenty minutes at ten pounds pressure (240°).

Drain the fish over a wire screen for several minutes, then pour two to four tablespoons of heated vegetable oil over the fish before closing the jars. Process the jars at ten pounds pressure for one hour and forty minutes.

Smoked Salmon

There are two methods for canning smoked salmon. One method is to fillet and cold-smoke the raw fish for fourteen to sixteen hours before cutting them to jar length and packing them. The second method is to fillet the fish and cut them to jar length before smoking them for the same length of time. Precutting the fish increases the surface area exposed to the smoke and thus gives a greater smoke penetration for a more intense flavor. In addition, it causes the product to be drier and to have a deeper color.

Pack the fish firmly in pint jars, leaving a one-inch head space. Add one and a half teaspoons of salt to each jar. If you use the second method, pour two tablespoons of heated oil over the top before closing the lid.

Process the jars for two hours at ten pounds pressure (240°).

Smoked Clams and Oysters

Pack the smoked clam or oyster meats tightly in pint jars, leaving half an inch of head space. Brush the top of the meat with vegetable oil and close the jars. Process the jars for two hours at ten pounds of pressure (240°).

5

Pickling

Pickling is one of the oldest, most effective methods of preserving fish, and it is a simple process that requires no special equipment. Pickling preserves foods because molds, yeasts, and bacteria cannot live in an acidic environment. A fifteen-percent acid content is necessary to preserve fish completely, but a content of only three percent is adequate if they are kept chilled at 40°. Curing in salt aids in the preservation process by drawing water, osmotically, from the fish. A moisture content of less than thirty-five percent has a pronounced effect on the ability of bacteria to sustain themselves. Salt firms the flesh of the fish, whereas acid (vinegar) softens it.

Food can be acidified in two ways. In fermentation the food reacts chemically in a relatively low-salt environment to change sugars and other compounds to lactic and other organic acids. At a certain point, the pH becomes low enough so that the bacteria can no longer grow. This is what happens when herring is allowed to "ripen" in recipes such as Gabelbissen and Matjesherring.

A second, simpler method is to pour vinegar of a strong enough acid concentration on the fish to reduce the pH to a level that inhibits bacterial action. A diluted acetic acid solution will have the same effect as vinegar. During the pickling process, any mold that forms must be removed. Not only can the molds produce disease-causing mycotoxins, but they also consume the bacteria-inhibiting acid.

It is essential to use only high-quality ingredients in pickling. The fish should be fresh, firm, and clean. Oily fish makes the best product, but even with these, care must be taken to assure a good pack. While herring is the most common pickled fish, only in the late summer and fall months does it have enough oil to be high in quality. Mackerel and the bellies of salmon (either fresh or salted) are also good pickled because of their high oil content. Salted fish can be pickled if they are sufficiently soaked beforehand.

Know the breeding season of the fish that you are pickling. A spawned-out fish, or a fish that is about to spawn, will not have enough oil or meat that is firm enough to make a good product. Salmon that are too close to the spawning season have thin muscle walls near the belly and the meat is inferior in quality.

PICKLING INGREDIENTS

The main ingredients in the pickling process are vinegar, water, salt, and spices. While there are an abundance of vinegars in common household use, the only one recommended for pickling is white distilled vinegar of from four- to six-percent acidity. Fruit and cider vinegars should not be used because their acid concentrations vary, and the fruit esters in the vinegar can sometimes give an unpleasant taste to the product. The darkness of these vinegars is also unattractive in pickled fish.

The acetic acid in vinegar is relatively volatile. If it is boiled too hard, the acid evaporates and the solution loses its preservative qualities. Whenever a recipe calls for a pickling solution to be boiled, bring it to a boil and reduce the heat to just below boiling for the remainder of the cooking time.

Hard water, especially water that contains large amounts of iron, magnesium, or calcium, colors the pickling solution and gives it an undesirable taste. If you do not have a water softener, boil the water for fifteen minutes to soften it. Skim off any scum that forms on the surface and let the water sit for twenty-four hours.

Use fresh whole spices in making pickled fish; ground spices tend to cloud the pickling liquid. For small batches of fish, you can buy commercial pickling-spice mixtures rather than mixing the individual ingredients.

In recipes calling for garlic, press the cloves and let them simmer with the other spices. Garlic is highly susceptible to bacterial action (notably *Clostridium botulinum*), so always remove them from any spices that are included in the jars of pickled fish.

Many types of salt are available for home use, but the one recommended for pickling is a food-grade salt called canner's salt, or vacuum canner's salt. Other salts have additives—sodium silico aluminate and dextrose—to keep them free flowing and the iodine stable. Iodine (potassium iodide) is sometimes added to salt, and it will give a bitter flavor to fish and cloud the pickling liquid.

Salt can be obtained in either a ground form or in flakes. The different types are equal to each other in weight. But in volume measurement, one cup of ground salt equals one and a half cups of flake salt. A pound of granulated salt is approximately one and a half cups—meaning that a cup of granulated salt weighs about ten ounces.

When vinegar and salt are combined they can corrode metals. Zinc, iron, copper, brass, and galvanized metal containers should never be used for pickling. Glass, stainless steel, aluminum, unchipped enameled steel, and paraffin-lined wooden kegs are all suitable pickling containers. Crocks and stone jars also work well. Whatever container you choose, clean it thoroughly first.

The recipes in this chapter are of two types: those that require cooking the fish and those that don't. Both methods produce satisfactory results and have equal storage capability. Pickled fish keep very well, depending on the conditions of storage, and they taste best after they have aged for a few days.

Keeping the pickled fish at low temperatures in a dark place will prolong their shelf life for months, but even at room temperature the pickling process will preserve the fish for up to two weeks. For longer storage times, or for a more pronounced pickle flavor, let the fish age for ten days to two weeks and then pour off the pickling fluids. Make a new solution and pack it with the fish.

Always label the product, noting the date of the pack and the process used. Once you have tried the basic recipes, experiment with other spices and ingredients. Use chili peppers, cayenne, allspice, nutmeg, cinnamon, cardamom, coriander, and any other spice that might give fish a good flavor. Try other types of fish than those listed. You can substitute dry white wine or burgundy for up to twenty-five percent of the vinegar in the sauce.

PICKLING RECIPES

Salted Salmon

The simplest method for pickling salmon is to use salted fish. Freshen the salted fish by soaking the pieces in cold water (ice can be added to cool it) that is running at a slow trickle. Four to eight hours of soaking are required, depending on how hard the fish are cured.

Make a vinegar sauce by combining equal parts of distilled vinegar and water. Add a small amount of sugar to take some of the edge off the vinegar taste and to give a sweet-sour effect to the liquid.

Wrap commercial pickling spice (available in most grocery stores) in cheesecloth and suspend it in the vinegar sauce while bringing the liquid to a boil. Simmer the vinegar and spices for ten minutes. Instead of buying commercial premixed spice, mix your own spices and put them in the cheesecloth bag.

Cut the fish into one-and-a-half-inch cubes and pack them in sterilized pint jars. Place onion slices between every other layer of fish. Pour the still-hot sauce over the fish and close the jar.

Fresh Salmon, Precooked

If you are using fresh salmon, thoroughly wash, fillet, and bone the fish. Cut it into two-inch pieces and dredge them in salt. Put them aside to let the salt absorb some of the water from the tissue.

After thirty minutes, rinse the salt from the fish and cook the pieces in slowly boiling water until they are firm but *not* over-cooked.

While the fish are cooking, slice an onion or two and sauté the rings in one-half cup cooking oil until the onions are trans-parent. Add three or four crushed bay leaves and six or eight whole cloves to the onions. A tablespoon of black peppercorns and a tablespoon of mustard seed can also be added. If desired substitute three or four tablespoons of commercial pickling spice for the individual ingredients.

Make a half-and-half solution of distilled vinegar and water and pour it over the onion-spice mixture. Simmer the sauce for thirty minutes.

After the fish pieces are cooked, cool them in the refrigerator and then put them in a sterilized container with the pickling sauce. Put the container in a cool, dark place for two or three days to age.

At the end of that time repack the fish into sterilized pint jars. Strain the pickling sauce through cheesecloth and pour the liquid over the fish. If the liquid tastes too spicy, thin it with more vinegar and water in a fifty-fifty ratio, or add a little sugar. A small amount of fresh spice can also be put in the jars before they are sealed.

The size of the fish pieces, the amount of the spices and their strength, and the time the fish age in the container all determine the flavor of the fish, so experiment to find the ideal recipe. Al-ways label the jars so that you can repeat it if it's a good recipe or discard it if it isn't.

Fresh Salmon, Uncooked

To pickle fresh salmon without cooking it, clean and fillet it and cut it into serving-size pieces. Make a brine of three-quarters of a pound of salt (about 1¼ cups) per gallon of water and combine the mixture with a gallon of distilled vinegar. Soak the fish pieces in the vinegar brine for three or four days, agitating them periodically to keep the salt dissolved. Remove the fish from the soaking solution and pack the pieces into sterilized pint jars. When packing, alternate layers of sliced onion, pickling spice, and fish. Cover the fish with a fresh vinegar brine and store the sealed jars in the refrigerator. Let the fish cure for two days before eating them.

Mackerel and Shad

Fillet thoroughly cleaned mackerel or shad, making sure none of the backbone remains. Cut the fillets into serving-size pieces. Fill a container halfway with salt and drop the pieces in. Mix the pieces and salt until the pieces are covered. Put them in a separate container, allowing the salt that sticks to the fish to remain. Let the fish sit in the container for two hours while preparing the pickling liquid.

Add two chopped onions to a vinegar-water solution (4 cups vinegar to 3 cups water) and bring it to a boil. Press the juice from a clove of garlic into the water. Tie in cheesecloth a mixture of equal portions of allspice, cloves, black peppercorns, bay leaves, and nutmeg and add it to the boiling liquid. Five or six tablespoons of commercial pickling spice can be substituted for the individual spices.

Simmer the mixture gently for ten minutes while rinsing the salt off the fish pieces. Add the fish to the spice sauce and bring it just to the boiling point. Simmer the pieces for ten minutes. Remove the fish and cheesecloth and strain the liquid before bringing it to a boil again. Add one tablespoon of sugar or sugar to taste. Pack the fish in sterilized pint jars and drop in a few pieces of fresh spice, or add a bay leaf and an onion slice to each jar. Fill the jars with the boiling liquid and seal them immediately.

Escabeche: Mackerel, King Mackerel, Red Snapper, Tuna, and Mullets

Any firm-fleshed fish can be pickled by the Spanish process called *escabeche.* Cut the thoroughly cleaned, fresh fish into serving-size pieces (about two inches long) and soak them in a 100-percent brine for 30 minutes.

Chop and mix together several bay leaves, one clove of garlic, and some dried red chili peppers and put them in a frying pan filled to a depth of one inch with olive oil.

Wipe the fish dry and fry them for thirty seconds, or until they are a light brown around the edges. Remove the fish and put them in the refrigerator to cool.

Add sliced onions to the oil and cook them until they are clear and soft. Pour 4 cups of vinegar into the oil-onion mix and then add ½ tablespoon each of cumin seed, marjoram, thyme, and black peppercorns. Cook the mixture slowly for thirty minutes and allow it to cool.

Pack the cold, fried fish pieces into sterilized pint jars and fill the jars with the sauce. A small amount of fresh spices can be added before sealing the jars. Store the pickled fish in the refrigerator.

Boiled and Pickled Fish

An alternative to frying the fish is to boil them. Large fish work best. In addition to mackerel, snapper, tuna, and mullet, you can use haddock, hake, halibut, and other white-fleshed fish.

Thoroughly clean and skin the fish and cut them into serving-size fillets. Make a 100-percent brine and soak the fish for three to four hours. The fish can be soaked for a longer or shorter period, depending on how fast they absorb the salt and how salty you want them to be.

After brining the fish, rinse them briefly in fresh water and place them in a large enameled pot. Layer the fish between sliced onions and a pickling mixture of bay leaves, mustard seed, red chili peppers, white pepper, allspice, and cloves.

Cover the fish with 2 quarts of six-percent vinegar mixed with 1 quart of water. Bring the mixture to a boil and reduce the temperature to simmer. Cook the pieces until they are easily pierced with a fork.

Remove the fish from the pickling broth and cool the pieces. Strain the vinegar-spice solution and bring it again to a boil. Pack the fish in sterilized pint jars, adding a few fresh spices and a lemon slice to each jar. Pour the heated liquid over the fish and seal the jars. Store the pickled fish in the refrigerator.

Haddock, Hake, and Halibut

To pickle haddock, hake, and halibut without cooking, a two-stage process is used. Soak the cleaned and filleted fish for two days in a brine solution of one pound of salt (1½ cups granulated) and one gallon of water. Keep the temperature at 40°. The fish should be quite firm at the end of this time.

Repeat the same soaking procedure for the same amount of time using five-percent distilled vinegar instead of salt water. Keep the vinegar cold. For an average-size piece of fish, it takes up to seven days for the acid to penetrate to the center of the meat, while it takes only twenty-four hours for *botulinum* spores to begin growing if conditions are favorable. It is imperative, then, to keep the fish cool to prevent the formation of toxins.

Prepare a pickling sauce by combining 1 cup of sugar for every 2 cups of distilled vinegar and bring it to a boil. Wrap some pickling spices in cheesecloth and suspend them in the boiling solution for five minutes. Remove the sauce from the heat and let it cool.

Alternate layers of fish and thinly sliced onions in sterilized pint jars, dropping a few pieces of fresh spice in, and pour the cooled vinegar sauce over the fish. Marinate the fish for several days in the refrigerator before eating them.

Herring

Herring can be pickled in two ways. The first—and easiest—method is to dry-salt the fish and pack them in buckets, crocks, or kegs. Gib the fish and pack them with the bellies facing up, heads to the outside, and the tails overlapping. Or clean and fillet the fish before salting them. The fillets need not be arranged in any particular order as they are salted. But use enough salt in packing—approximately one-third of a pound of salt for each pound of fish—to guarantee a 90° hydrometer reading for the brine that forms.

After the fish have cured for six to eight weeks at a cool temperature (preferably 45°–55°), remove them from the container and rinse them in cold water for four to eight hours. The soaking time depends on the thickness of the fish, the rate of water flow, and the length of time in the salt cure. The fish should still have a slightly salty taste when they are removed from the fresh-water soak. From this point, they are pickled according to the recipe given for salted salmon in chapter 7.

The herring can be cut in a variety of ways. The most convenient is to fillet the fish and cut them into inch-and-a-half pieces.

Or the herring can be cut perpendicular to the backbone in one- to two-inch pieces, leaving the backbone in.

Rollmops can be made by removing the backbone from the pieces cut in this way. The meat is then wrapped around a piece of dill pickle. The rolled fish can be held closed with either a toothpick or a whole clove. *Bismarck herring* is like *rollmops* without the pickle. The fish is boned without cutting the backbone similar to the butterfly cut described for black cod. Pack the fish in glass jars with a slice of lemon and cover them with the pickling liquid.

The second method for pickling herring, although more complicated, produces a finer pack. Clean, scale, and cut the heads off the fish. Wash the belly cavity and scrape the kidney from the backbone. Mix one part of five-percent distilled vinegar with an equal part of fresh water and add salt to make a brine that measures approximately 90° on the hydrometer. Soak the fish in the brine for five days, keeping them cold. Then repack them in a vinegar-water brine in a 4:3 ratio. Add enough salt to this three-percent vinegar solution (about 13 ounces per gallon) to obtain a 35° reading on the hydrometer. Store the fish in the refrigerator in the brine until they are ready to be pickled.

Make a simple pickled herring from these vinegar-brined fish by soaking one-and-a-half-inch pieces in fresh water for eight hours. Slice an onion and layer it with the fish in pint jars, alternating onion and fish. Sprinkle one or two tablespoons of pickling spice (any mixture of black peppercorns, bay leaves, mustard seed, allspice, cloves, and red peppers will do) throughout the onions and fish. Again, prepare a four-to-three vinegar-water brine and bring it to a boil. Add enough sugar to give a sweet taste (start with three tablespoons per gallon of solution) and pour the hot liquid over the fish before sealing the jars. Store the jars in the refrigerator and age the fish for ten days before eating them.

There are many variations on the basic herring recipes given above. A sour-cream dill-weed sauce or wine sauce can be made for the pickled fish. Try different combinations of spices and sauces with these basic recipes. Experimentation and good records are the key to successful herring pickling. The most important factor to remember in the pickling process is the acid. It is responsible for preserving the fish, and its concentration must be at least three percent.

Sturgeon, Pike, and Pickerel

Wash the fish thoroughly and cut them into individual serving-size pieces. Dredge the fish in a fine-ground salt and let them sit for three hours.

While the pieces cure, mix a solution of one quart white wine, one quart distilled vinegar, and one pint of water.

Rinse the salt from the fish and pat them dry. Brush the fish pieces with vegetable oil and broil them or fry them. When the fish have turned a light brown, remove them from the heat and let them cool.

Place the fish in clean glass jars, alternating layers of fish with sliced onion, whole cloves, and whole black peppercorns and rosemary. Thyme can be substituted for the rosemary. Pour the sauce over the fish and seal the jars.

Clams and Mussels

Scrub the shells of the clams or mussels and steam them briefly to open the shells. Discard any of the bivalves whose shells open before they are steamed. Open shells indicate that the clam or mussel is dead and probably spoiled. Save the nectar from the steaming to use later.

Remove the meats from the shells. If you are using mussels, trim the *byssus,* the fibrous filaments with which they attach themselves to pilings and rocks. Cool the meats and layer them with chopped onion, a bay leaf, and whole cloves in sterilized pint jars. Place a towel over the jars to keep the light and air from turning the meats a dark color.

Strain the steaming liquid and add it to distilled white vinegar in a four-to-one ratio of liquid to vinegar. Wrap pickling spices in a cheesecloth and suspend them in the liquid. Bring it to a boil and reduce the heat to a simmer. Cook the mixture for forty-five minutes and pour it over the meats in the jars.

Because the acid concentration is not very high in this recipe, the jars must be stored in the refrigerator. Let the pickles age for a couple of weeks before consuming them.

Oysters

Oysters can be pickled by the method given for clams and mussels, or according to the following recipe.

Open the oysters and save the liquor. Strain the liquor to remove shell bits and other debris and add it to salted water. The old saying about the salted water is that it should be the same concentration as the seawater from which the oysters came, or about ten percent.

Pour the resulting liquid into a large enameled pan and bring it to the boiling point. Drop the oysters in one by one and blanch them until the edges begin to curl. Remove the cooked meats and set them aside to cool.

Use the blanching liquid to make a vinegar sauce of two parts liquid to one part distilled white vinegar. Bring it to a boil, then reduce the heat to simmer.

Wrap commercial pickling spices in a cheesecloth or make a mix of fennel, allspice, bay leaves, crushed garlic cloves, thyme, black peppercorns, mace, cinnamon, and cloves. Dangle the spices in the simmering sauce and let them cook together for forty-five minutes. While the sauce cools, pack the oysters in sterilized pint jars. Add a slice of lemon to each jar and pour the sauce over the meats. Seal the jars and store them in the refrigerator.

6

Drying

Drying is one of the oldest methods of preserving fish. Long before refrigeration and freezing were common, drying (as well as salting) provided a means of keeping fish for extended periods. Drying, or dehydrating, can be done using fresh or salted fish.

Whether the fish are salted or fresh, the desiccation process involves many steps that can be roughly grouped into two major divisions. The first is called the *constant-rate phase.* The second is the *declining-rate phase.* In the constant-rate stage of drying, the rate of dehydration occurs at a steady pace. This stage lasts from six to ten hours.

Water is removed from the tissue by simple evaporation. As anyone who has worn a wet shirt when the wind is blowing can testify, evaporation produces a cooling effect. This same situation exists with fish: The surface of the tissue cools as the water evaporates and the air immediately next to it becomes higher in moisture content. Dry air moves in to replace the moister air and the process is repeated. This action causes minor fluctuations in the speed of drying so that the term *constant-rate* is really just an approximation.

During the declining-rate phase of the drying process, the surface of the fish is dry and simple evaporation no longer accounts for the rate of drying. The limiting factor here is the rate that water can travel from the interior of the fish—through capillary action—to the surface. Capillary action is much slower than evaporation, and it occurs at a steady rate that can be neither speeded up nor slowed down. The rate of drying declines because the moisture from the interior of the fish takes longer to reach the surface, even though it is traveling at a steady rate.

In the past, cod was the most commonly dried fish, and therefore, it has been the most thoroughly studied. The water content of cod is around eighty percent: That is, for every ten pounds of cod muscle, eight pounds will be water. These figures are approximately the same for all lean species (whitefish) of ocean fish. Because fatter fish have a higher oil content and consequently a lower water content, they do not dry as well. The oil in the flesh has a tendency to become rancid.

In order to achieve a state where no molds or bacteria can grow on the fish tissue, the moisture content of unsalted fish must be lowered to sixteen percent. For fish that have been salted, the moisture content must be brought down to thirty-three percent of its initial weight.

The first fish ever dried were most likely hung in the open air during mild weather. The combination of factors necessary to dry fish successfully—a readily available supply of fresh fish of good quality, warm but not hot temperatures, weak sunlight, and the absence of insects—were not present in many areas during much of the year. One place that has the proper climate is a small area north of Ålesund, Norway, where cod is hung in the spring (before flies became a problem) in the open air to dry. Drying requires two to six weeks to drop the moisture content to the required sixteen percent.

AIR TEMPERATURE AND HUMIDITY

Because the combination of factors required to air-dry fish properly were so seldom right, artificial means were devised. Ovens and drying tunnels were designed to simulate optimal drying conditions. Many experiments and much research has determined that the best conditions for fish drying occur when the air temperature is 75° (or from 60° to 80°) and the relative humidity is forty-five to fifty percent. A steady breeze of four miles an hour is also needed.

Temperature plays a big part in desiccating fish, especially because it helps to determine the length of the constant-rate period of evaporation. The rate of evaporation is independent of the air temperature, except that the air temperature affects the temperature of the fish muscle. High temperatures must be avoided because they increase the rate of oxidative rancidity in oily fish.

The relative humidity of the air surrounding the drying fish is also a critical factor: The drier the air, the faster the fish dries. The air temperature determines the quantity of moisture that the air can hold. Higher-temperature air can absorb more water from its surroundings and thus isn't able to absorb much moisture from the fish.

If the relative humidity is too high, the atmosphere won't be able to absorb water from the fish. If it is too low, the surface of the fish will dry out at a faster rate than the interior and becomes overly dry by the time the interior is dry. Low relative humidity—near 45 percent—is best when the surface of the fish is wet (during the first phase of drying), but it is harmful during the dry-surface stage. When the exposed flesh of the fish is dry, the relative humidity has little impact on the rate that moisture is able to travel through the meat itself.

As a corollary to the relative humidity, the dew point also plays a part in drying fish. The dew point is the temperature at which the humidity in the air begins to condense to form a liquid. A high relative humidity has a high dew point. If there is much moisture in the air, the temperature does not have to drop very far before the moisture begins to condense. To avoid condensation the temperature must be kept higher than the dew point. Assume that the maximum air temperature for drying fish is 80°. Fish will always dry when the dew point is below 55°, but it will dry much slower at 60°. When the dew point reaches 65°, the fish will not dry.

The air velocity at the surface of the fish is also important. When water evaporates and cools the fish, the shifting air moves in to rewarm the surface by convection. The air immediately next to the surface of the fish has a higher relative humidity than the surrounding air because of the water leaving the tissue. It may even become saturated to the point where it can no longer absorb water from the fish. Then, air movement replaces the wet air with drier air to continue the cycle.

The velocity of the air has much the same impact on the drying fish as does the relative humidity. Good air movement during the wet-surface stage is important in promoting drying, but too much—during the dry-surface stage especially—will cause overdrying. It is often just as easy to inspect the fish during drying and to adjust the air flow as it is to establish a particular air velocity for the entire dehydration process.

PRESS PILING

Thin fish fillets, strips, or flaked fish can all be dried in a continuous process. One temperature, a single relative humidity, and a steady air flow will ensure uniform drying because of the uniformity of the fish pieces.

Thicker fillets or pieces of fish with varying thicknesses require a procedure of discontinuous drying called *press piling.* In the press-piling process, the fish are dried until the surface of the flesh is dry. Then they are removed from the dehydrator and pressed until water works its way up from the interior of the fish to wet the surface again. If a large number of fish are involved, they are simply piled on top of each other to press the water out. Of course the fish on the bottom have more pressure on them and lose more water, so to keep the water content of the fish the same the fish should be repiled halfway through the process by putting the top fish on the bottom and the bottom fish on the top.

After they have been pressed, dry the fish again until the surface has lost all its moisture. If necessary, press the fish a second time and dry them again. Pressing the fish smoothes the surface as it dries and results in a more attractive product.

A word of caution regarding press piling: If the fish are stacked on top of each other and allowed to sit for an extended period when the surfaces are wet, slime-forming bacteria will grow. To control these bacteria, dry the fish again as soon as the surface is wet. Sliming can also be a problem during the first stages of drying. High temperature, high relative humidity, and poor air circulation also can cause sliming.

STOCKFISH: DEHYDRATING FRESH FISH

There are several ways in which fresh fish can be dehydrated. When fish is dried without any other treatment it is called *stockfish*. The product is hard and dense, with a low water content. The flesh of the fish shrinks considerably during the process.

Air drying is perhaps the simplest way to reduce the water level of fish. To dry lean, white-fleshed fish, begin by thoroughly cleaning and dressing the fish. Split large fish into fillets or cut them into one-and-a-quarter-inch-wide strips. By cutting the fillets into strips, the surface area is increased and the fish will dry faster. (See chapter 2 for an easy way to hang strips.)

Alternatively, remove the backbone from the fish as described for black cod in chapter 2 to make a butterfly fillet. Hold it open as the fish dries by placing sticks on either side and binding them to keep the fish flat. Shift the location of the sticks to avoid wet spots. A butterfly cut can also be scored through the flesh—but not through the skin—to form "strips" that are still attached to the skin.

After dressing the fish, soak them in a brine solution of one cup salt per gallon of water for ten minutes. This brief brining will add a small amount of salt to the surface of the fish and inhibit microbial action as the fish dries.

Hang the brined fish in any of the ways discussed for hanging fish to be smoked. (See chapter 2.) Use "S" hooks or racks, or tie two fillets at the tails and hang them over sticks with the skin sides facing toward each other. Fold butterfly cuts in half, with the flesh side out, and drape them over horizontal sticks to dry.

Put the fish in a shady spot where there is a steady breeze. Direct sunlight will cook the fish, especially during the dry-surface stage of drying. If the fillets are thick, or if the fish are drying slowly, cover them at night to prevent dew from condensing on the surfaces and rewetting them. It may be easier to bring them indoors. If insects are a problem, cover the fish with cheesecloth. The cloth will keep the insects out but allow water vapor to escape.

It was once common among Norwegian fishermen to hang strips of cod or flounder in the rigging of their fishing boats. The dried strips, called *rekling,* were brined as described above and hung for several weeks to dry until they were hard.

DRYING FLAKED FISH OR FISH PIECES

To dry flaked fish or fish pieces, place them on cookie sheets or fine-meshed screen in the open air.

The fish don't have to be soaked in plain brine before they are dried. The brine can be flavored with any of the spices for the pickling brines in chapter 5. The recipes for the pickling solutions in chapter 5 make good marinades for fish that are to be dried, but don't keep them in the brine for more than ten minutes. Otherwise the vinegar in the pickle will begin to soften the flesh. Remember that any spice flavor will become more concentrated as the fish dry, so don't overpower the flavor of the fish with spices.

Keep the salt concentration to about one cup of salt per gallon of water. Because they are small the pieces have a large surface area compared to their size and absorb salt quickly. Be careful not to overbrine the fish. The hanging or racking procedure, as well as the drying time, is the same for flavored fish and brined fish.

If weather conditions are not suitable for drying fish outdoors, you will have to resort to mechanical means. A home oven makes a good dehydrator, but be careful not to *cook* the fish. Prepare the fish as though you were going to dry them outside. After brining, hang the fish from the oven racks or simply lay them perpendicular to the racks and let them dry for several hours. If the pieces are small or flaked, place a piece of fine-meshed screen over the oven racks and put the fish on it. A pellicle will form on the surface during the first few hours of air drying.

Once the pellicle has formed, turn the oven to its lowest setting to bring the temperature up to 80°. Since most ovens do not have a setting for 80°, leave the oven door open to prevent the air from becoming too hot. With the door open, air circulation will be increased and drying will proceed more rapidly. Remember, the idea is to dehydrate the fish slowly, over a period of time, not to cook them.

Another way to dry fish is to hang them in the opening of a fireplace. The draft of the chimney will provide the necessary air circulation (be sure to open the draft before hanging the fish to avoid dislodging soot particles). To raise the temperature around the fish, place a hot plate under them. Do not place the hot plate where fish oil can drip on it and cause a fire, though! Rearrange the fish periodically to guarantee uniform drying.

FOOD DEHYDRATOR

The following plans for a food dehydrator are adapted from information provided by the Extension Service, Oregon State University. Professor Dale Kirk, Agricultural Engineer, wrote the initial instructions.

The dehydrator shown in Fig. 31 can be built from ½ inch plywood in just a few hours. In addition to a 4 feet by 8 feet by ½ inch sheet of A–C exterior-grade plywood, the following materials will be needed:

9 ¾″ x ¾″ x 4′ wood strips
1 pr. 2″ metal butt hinges
100 1″ #8 flat head wood screws*
1 ball chain door latch
18 ⅝″ #7 round head wood or sheet metal screws
    (for the light bulb sockets)
5 aluminum screens
    16¾″ x 20″
    16¾″ x 19″
    16¾″ x 17¾″
    16¾″ x 16¾″
    16¾″ x 15½″
1 6″ or 8″ fan
9 porcelain surface mount sockets with concealed terminals
9 75-watt incandescent light bulbs
15′ #14 asbestos-covered copper wire
6′ #14 extension cord with male plug
1 thermostat, 10-amp capacity, 50°–100° range with remote
    sensing bulb
1 oven thermometer that reads from 50° up
1 4″ electrical surface utility box with blank cover
2 ½″ utility box compression fittings
2 wire nuts, 3 #14 wire capacity
1 36″ x 14″ sheet heavy-duty aluminum foil wrap

*1 ⅛″ brads and glue may be substituted

The materials listed above include five aluminum window screens that are used as drying trays. Of course, it is possible to make the screens at home, but it is much simpler to buy the trays at a lumber store or building supply dealer. To make the trays yourself, simply build frames out of light wood stock to the dimensions given for the window screens. Use either aluminum screen or hardware cloth to hold the fish. Try not to use plastic screens as they will sag under the weight of the wet fish. Sagging

Fig. 31. A homemade dehydrator: Nine light bulbs mounted on the floor supply heat; a thermostat controls the temperature, regulating it from 0° to 150°; the fan provides air circulation.

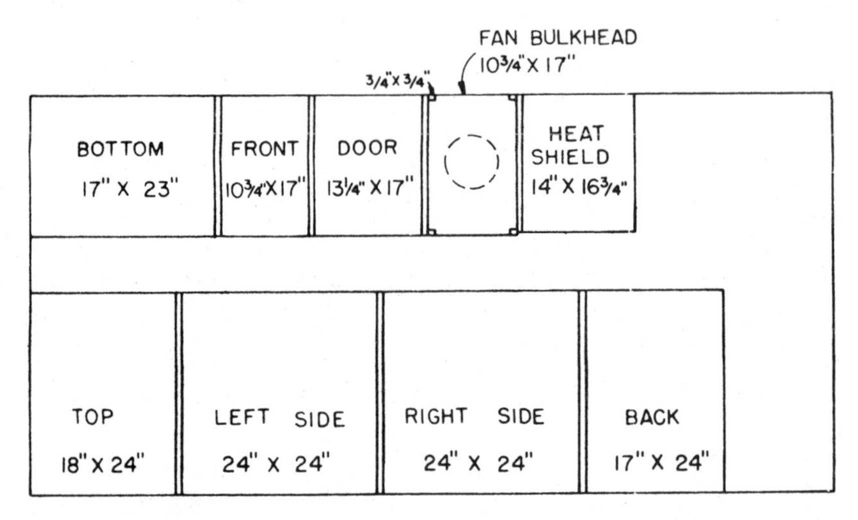

CUTTING PLAN FOR PLYWOOD PARTS SHOWING ACTUAL DIMENSIONS. ALLOW FOR SAW KERFS ON SAWN EDGES.

Fig. 31A.

can be reduced, however, by placing light wood strips under the screen material to serve as braces.

Because the range of temperatures for drying fish is close to room temperature the easiest place to find a thermostat is a home heating equipment dealer. An appliance parts dealer might also have the proper thermostat. The problem will be in finding a thermostat with a remote sensing bulb.

The nine light bulbs in the dehydrator are capable of raising the temperature to over 150° in the area close to the bulbs. By placing the thermostat (or the sensing bulb) in different locations within the dehydrator the temperature at which the thermostat will cut the power to the light bulbs will change. For example, a thermostat located at the upper left corner of the back wall may be set to turn the light bulbs off when the temperature reaches 80°. When the lights go off the temperature at the upper front of the dehydrator (as measured by the thermometer) may only be 75°. This is not a significant difference but it illustrates the fact that the location of the thermostat will affect the temperature in different parts of the dehydrator.

You can use this fact to your advantage. Say that your thermostat functions in the 110°–150° range. This temperature is too hot to dry fish, but the thermostat can still be used. Just place it closer to the light bulbs. The thermostat will reach 110° faster than if it were farther away and will turn off the lights even though the temperature in the upper parts of the dehydrator is only 80°. For this reason, an ordinary meat thermometer or oven thermometer should be placed in the upper front portion of the food dryer to indicate the "average" temperature in the dehydrator.

A thermostat with a wider operating range will allow the dehydrator to be used for other foods besides fish.

Building the Dehydrator

Begin constructing the dehydrator by cutting out the pieces from the plywood. Figure A shows how to draw the dimensions on the plywood to minimize waste in cutting. When laying out the dimensions be sure to leave some space between pieces to allow for the wood cut out in the saw kerf.

Next, cut the ledgers, shown in Figure B, from the ¾ x ¾ inch stock. The lengths for these ledgers are as shown: 22″, 21⅜″, 20¾″, 20³⁄₁₆″, 19⁹⁄₁₆″, 19³⁄₁₆″. The diagonal strip of ¾″ material serves as a stop for the trays, preventing the racks from sliding into the sensing bulb of the thermostat. The four-inch space at the bottom of the diagonal also enables air to circulate up from the light bulb area.

Assemble the side panels as shown in Figure B. The dimensions for the spacing of the ledgers is given in Figure D. Note the sensing bulb hole for the thermostat. Depending on the particular thermostat, the sensing bulb may be an integral part of the unit so the entire instrument will be inside the dehydrator. In this case, no holes will be needed. The location shown on the plans is a sample only. Your thermostat may require different positioning. Since

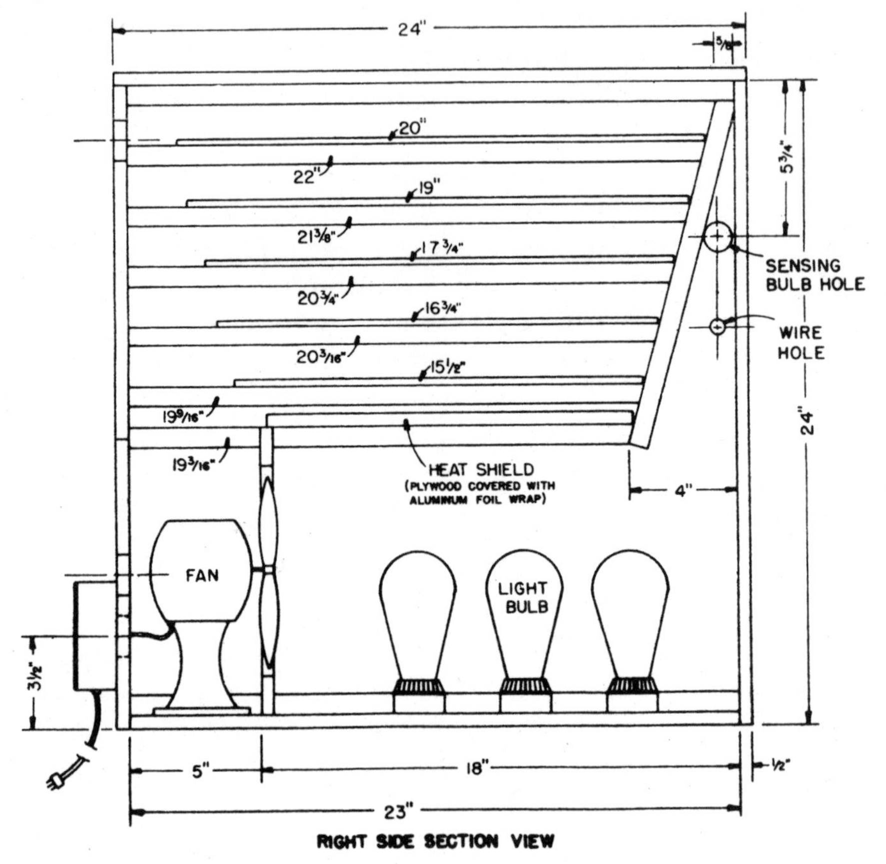

Fig. 31B.

there is only one thermostat, there is only one hole. Do not drill the holes in both side pieces.

Now, lay out the light sockets on the bottom piece of plywood. Figure C shows how to arrange them. Fasten the asbestos-covered wire to the sockets. Connect the wire that runs to the yellow screws on the sockets (the screws that connect to the center pole rather than the threaded wall of the socket) to the thermostat. Connect the wire that runs to the white screws to the white wire in the extension cord. Connect the third wire in the extension cord (the green one) directly to the junction box that is mounted on the front panel.

The next step is to attach the fan. A household fan with a base is the easiest type of fan to work with. Secure the base of the fan to the bottom of the dehydrator and cut a hole in the fan bulkhead to fit the diameter of the fan blades. Figure C shows the positions of the fan and the fan bulkhead. Figure B shows the distance between the bulkhead and the front piece to be 5″. This length can vary a little bit, but the fan motor should be no more than 1″ from the front.

Screw the left side piece of the dehydrator to the bottom piece and make a temporary connection of the bulkhead to the side by

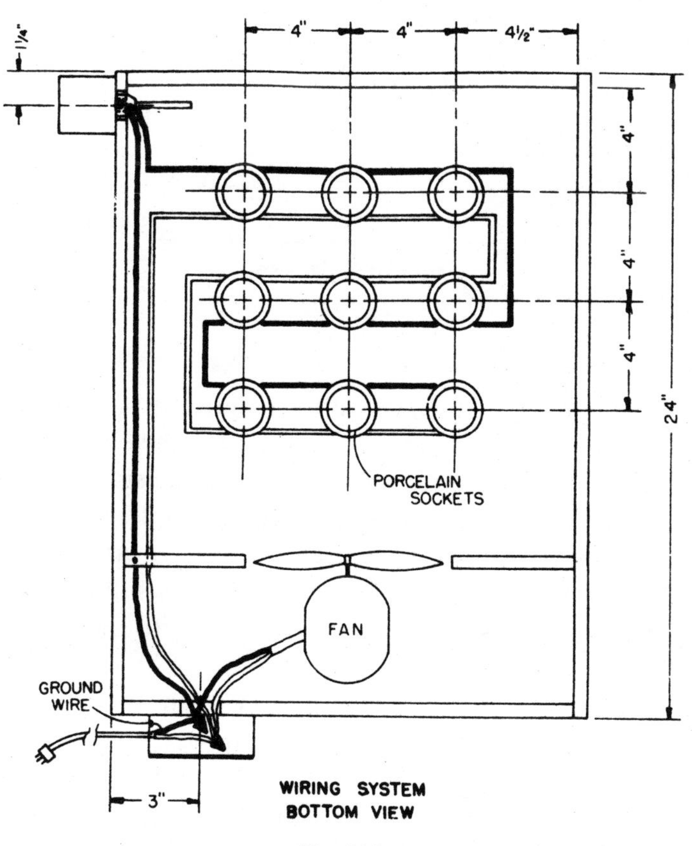

Fig. 31C.

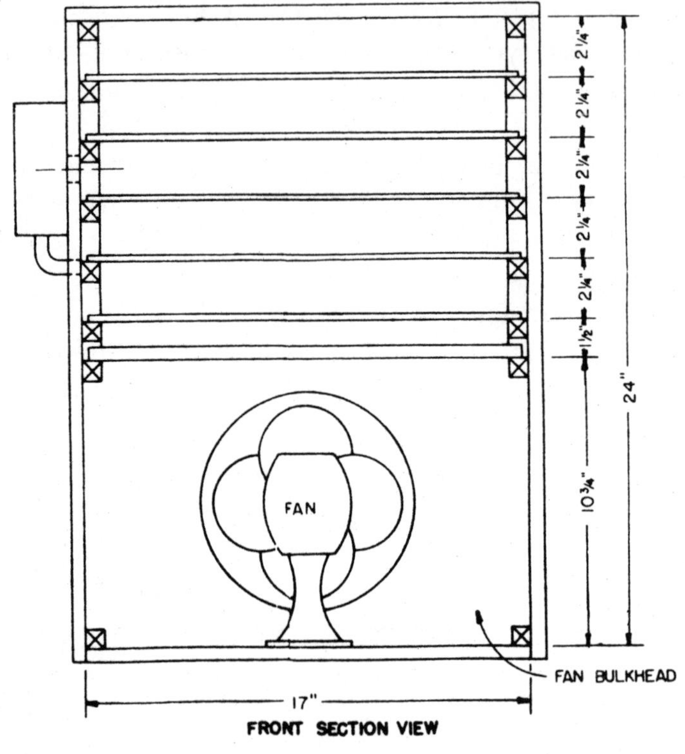

Fig. 31D.

placing two screws through the left side into the bulkhead. Screw the right side onto the bottom piece and onto the bulkhead. Also, attach the back and the top of the dryer to the sides.

Cut the 1½″ hole in the front piece so that it is centered right in front of the fan motor. Since the fan motor is within an inch of the opening all the cool air drawn in by the fan will pass over the motor and keep it cool. Secure the front piece to the bottom and the sides with screws.

Now, wrap the plywood heat shield with the aluminum foil so that the shiny side of the foil faces out. The heat shield reflects the heat from the light bulbs downward to prevent the bottom tray of fish from cooking. Lay the shield on its ledgers without screwing it down. Then, it will be easier to remove for cleaning.

At this point, all that remains is to attach the door. Position the hinges as shown in Figure E and drill the 1½″ diameter exhaust hole up near the top. Make a simple latch for the door by drilling a small hole in the door and another one in the side right by the first one. Run a ball-link chain through the two holes and use it to control the width of the opening.

Ventilating the Dehydrator

During the early stage of drying, the constant-rate period, evaporation plays the biggest role in the dehydration of the fish. A

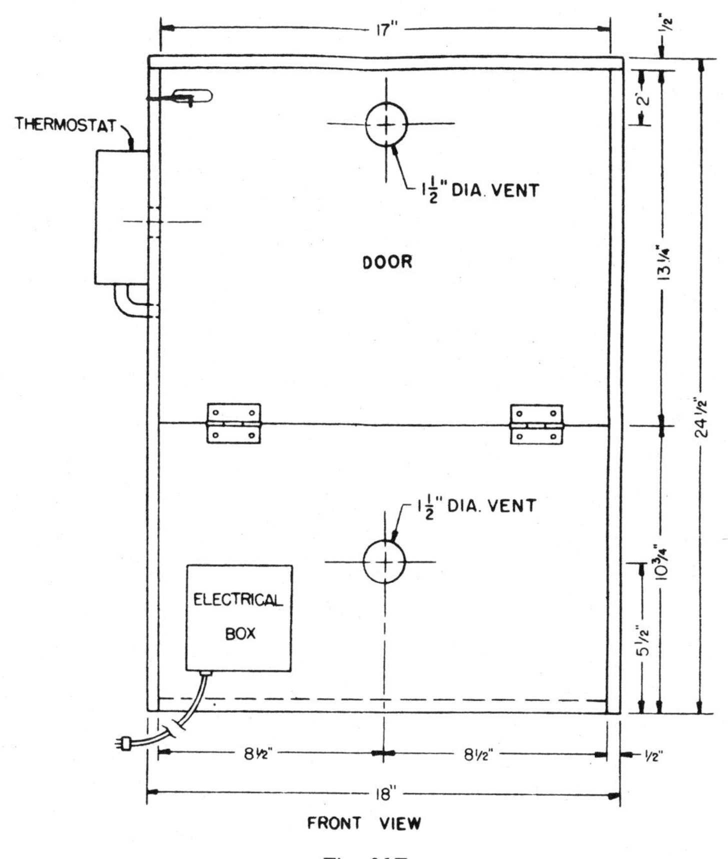

Fig. 31E.

large amount of water vapor will be leaving the tissue so it is best to use the ball-link chain to open the door about ¾".

You can check the stage of the drying by placing a mirror at the top of the opened door. When moisture no longer condenses on the mirror the constant-rate period is over. The door may then be closed. The air-intake hole at the bottom of the front piece and the exhaust hole on the door will allow enough air circulation to finish drying the fish.

Maintaining the Dehydrator

The dehydrator is easy to maintain. Clean the racks with hot water and detergent. Also, wash the heat shield to clean up any fish juices that may have dripped through the racks. Since the fan motor will be operating at an elevated temperature (because it is inside the dryer) its motor bearings must be lubricated regularly. Use a few drops of 30W engine oil. Light grades of household oil or sewing machine oil may tend to gum and will stall the fan motor.

DRYING SALTED FISH

Salted fish make a good dried product. Because the salt discourages bacteria and molds from growing on the flesh, it does not have to be dried as much as stockfish. The final weight of the dried fish should be thirty-three percent of the initial weight. *Saltfish* is the name given to fish that are salted before they are dried.

Salted fish is dried in much the same way as fresh fish. Because the salt has removed some of the water through osmotic pressure, the initial moisture content of the tissue is around sixty percent. Hang the fish and dry them under the conditions described for drying fresh fish.

The relative humidity of the surrounding air plays a more important role in making saltfish than it does in making stockfish. A low relative humidity causes rapid drying at the surface of the fish, which results in the meat becoming "case hardened." This tightening of the surface prevents the moisture in the interior of the fish from passing to the exterior. Counteract this tendency by partially rehydrating the fish before drying it. A low humidity level also causes lightly salted meat to crack and assume a rough appearance similar to that of stockfish. The coarseness of the surface can be moderated by press-piling the fish. In the winter months the relative humidity is often low. Use a small kettle of water and a hot plate to raise both the temperature and the relative humidity to higher levels.

REHYDRATING STOCKFISH AND SALTFISH

Stockfish can be eaten raw without rehydrating it, but it must be soaked out before it can be cooked. When it is dried, the stockfish tissue shrinks and becomes hard. It requires hours of soaking in a tub of cold water. The water must run quickly enough to completely replace itself every few hours.

The soaking time depends on the thickness of the fish, and it can vary from twelve to forty hours. Fish that have been pressed during drying will rehydrate slower than those that have not been pressed because the ability of the tissue to move water by capillary action has been disrupted. The water must be cold to reduce the level of bacterial activity. If the tap water is above 40°, use ice cubes to lower it.

Saltfish must also be soaked out if it is to be cooked. Since it has not lost as much moisture as the stockfish, its soaking time is not as long. It, too, must be soaked in cold water to increase the moisture content of the tissue, but it is also soaked to reduce the salty taste of the meat. Because of the structural changes that occur in the cells during the salting process, rehydrated saltfish has a chewy, somewhat rubbery texture and a cured taste. These characteristics can be masked in a variety of ways by the use of sauces.

RECIPES FOR DRIED FISH

Either saltfish or stockfish can be rehydrated and flaked in the following recipes. After soaking the fish, peel off the skin and remove any bones. Always save the last change of water used in the soak to cook any vegetables or potatoes called for in the recipe.

Lutefisk

Lutefisk, a traditional Christmas dish in Norway, is made from saltfish.

Soak dried or salted cod for up to three days to freshen it. Be sure the soaking water is cold and that it is changed often. When the fish have begun to fluff up and the salt taste has been reduced, drain it.

Soak the fish in a strong base (the *luting* phase). The Arm and Hammer Company makes a washing soda that will serve this purpose. It is composed of sal soda (hydrated sodium carbonate), a strongly alkaline salt used in bleaching and cleaning. It is irritating to the skin and eyes, so be careful when handling it. Mix the solution by dissolving a half-cup of the soda in a quart of hot water. Stir it continuously until the soda is completely dissolved. Any undissolved particles will "burn" their way into the flesh of the fish and lodge there. They *do not* taste good. When all the soda is in solution, add more hot water until you have a gallon of liquid. Stir it well to distribute the soda water evenly.

Put the fish in a wooden barrel or plastic bucket. Do not use aluminum because the soda solution will discolor the metal. When the soda solution is cool, pour it over the fish and let them soak for three days. Keep the temperature cool and do not change the fluid. Agitate the fish periodically.

At the end of the sixth day, remove the fish and put them in a container of cold water. Soak the fish in the water from four to six days, changing the water twice each day. After they are thoroughly soaked, the fish are ready to eat.

Put the fish in a covered container in the refrigerator or wrap the pieces tightly and freeze them. In the days before refrigeration, the fish were placed in a pail of fresh water and stored outside when the temperatures were near freezing.

It would seem that after all the soaking and bleaching that the nutritional value would be reduced, but lutefisk is a tradition, so it will probably continue to be eaten.

Prepare the lutefisk by removing the skin and bones. Place the fish on a rack in a saucepan or wrap it in cheesecloth before putting it in the pan. Cover it with cold water and add a little salt. Bring the water to a boil and reduce the heat to a slow boil. Cook the fish for twenty minutes and drain the water.

Serve lutefisk with boiled potatoes and have ready a thickened sauce to pour over the fish. The traditional sauce for lutefisk is a white sauce with ½ a teaspoon of prepared mustard per cup of sauce. In Sweden melted butter is served with the fish.

Fish Casserole

1 pound dried cod, hake, halibut, or other lean, white-fleshed fish (approximately 2 cups of flaked fish)
4 or 5 large potatoes
5 strips bacon
1 large onion, chopped
1 cup heavy cream
White pepper

Peel the potatoes and boil them in the fish water until they are half-cooked. Test them with a fork as they boil to avoid overcooking.

While the potatoes are cooking, flake the fish and set it to the side. Fry the bacon and drain it on paper towels. Sauté the onions in the bacon fat until they are soft and transparent. Pour off the excess fat.

When the potatoes are half-cooked, add the flaked fish to the potatoes and water and reduce the heat to a simmer. Crumble the bacon into the pan of onions and add the cream.

Warm the cream, but do not boil it. Drain the water from the fish and potatoes and put the food on a serving platter. Pour the cream, onions, and bacon over the top. Spice the dish with white pepper to taste.

Scalloped Fish

In this recipe, use dried cod or other fish and a variety of vegetables.

1 pound of dried fish, flaked
1 egg, beaten
2 cups milk
1 tablespoon flour
White pepper
1½ cups bread crumbs or crushed crackers
1½ cups chopped celery, sliced onions, or cooked lima beans
Parmesan cheese, grated
Butter or margarine

Place the fish in cold water with a bay leaf and a little thyme. Cold water will prevent overcooking the outer portion of the fish. Simmer the fish until it just begins to flake. Drain it and remove the bones and skin.

While the fish is cooking, add the egg to the milk. Mix in the flour and heat the mixture in a double boiler, stirring the sauce until it is thick. Add white pepper to taste.

Grease a baking dish with butter or margarine and put in one-third of the cooked flakes of fish. Sprinkle a layer of bread crumbs or cracker crumbs over the fish and add a layer of vegetables. Repeat the layering of fish, bread crumbs, vegetables, and finishing with a layer of bread crumbs on top. Sprinkle grated Parmesan cheese over the top of the casserole and dot the surface with butter.

Bake at 375° for twenty minutes, or until the crumbs are golden brown.

7

Salting

In the past, salting fish was a common method of preservation, but today salting is often merely the first step in a curing process. Salmon, cod, and herring, as well as many other fish can be preserved by packing them in salt.

Like drying and other methods of preserving fish, salting creates an environment in the fish tissue that is inhospitable to bacteria and other organisms. And like the heat process in canning, there is a balance in salting that must be achieved. Too little salt and the fish will be susceptible to decay from microbes. Too much salt and salt-loving bacteria called *halophiles* will spoil the fish. Too much salt also causes the proteins in fish to become denatured and inedible.

Sodium chloride is the predominant chemical in salt. At a concentration of one percent in the flesh of the fish, it actually stimulates the growth of certain bacteria. These bacteria die when the concentration reaches six to eight percent. Slime-forming bacteria are able to live in salt concentrations in the six-to-twelve-percent range if the temperature and relative humidity are high enough. The characteristics of sliming are a glistening, somewhat sticky yellow-gray layer of fluid between the fish and a sour smell to the brining liquid.

At concentrations above thirteen percent, red halophilic bacteria thrive. Red halophiles also produce a bad odor in the salted fish. The halophiles are so attracted to salt that if the relative humidity is high enough the salt will absorb moisture from the air and the bacteria will begin to grow. Always store salt in a cool, dry place to prevent contaminating the fish when salting them.

Dun is a condition caused by the bacteria *Sporendonema epizoum.* Small black spots grow on fish that are salted to a concentration of ten to fifteen percent if they are stored at 75° when the relative humidity is seventy-five percent.

The techniques discussed below will minimize the problems associated with these spoilage organisms.

Ordinary table salt is not pure sodium chloride. In addition to anti-caking compounds are chlorides such as calcium chloride and magnesium chloride. The magnesium and calcium also form sulfates. A yellow or brown color in salt indicates the presence of iron. The iron will stain fish either yellow or brown—although calcium chloride will help to remove the stain. Magnesium chloride retards the absorption of sodium chloride, the preservative agent in salt, and if it is present in a strong enough concentration, it will cause the fish to spoil because of bacterial growth.

The texture of salt also plays an important role in its use as a preservative. Coarse salt is less soluble than fine salt and takes longer to dissolve. This delay allows time for bacteria to grow and spoil the fish. Fine-grained salt, on the other hand, absorbs moisture from the air, making it difficult to sprinkle uniformly. It also allows the fish to pack too tightly, preventing the "pickle" from flowing between the fish. The resulting unsalted spots are susceptible to bacterial growth.

Fish can be salted in one of two ways. The first method, called dry salting, or pickle salting, is the most useful. Dry salting is especially good for fish that is later to be pickled, dried, or smoked. Brining, the second salting method, is not recommended for fish that are to be dried because it causes the thin parts to become saltier than the thick parts as water leaves the tissue.

THE DRY-SALTING METHOD

Large fish are filleted before they are dry salted; otherwise they are too thick for the salt to penetrate. Smaller fish can be cut open, cleaned and salted without any further treatment.

Begin the salting process by partially filling a small box with salt and dredging the pieces of fish through it, letting all the salt that sticks to the fish remain. It isn't necessary to rub the salt into the flesh of the fish, but be sure the salt covers both the inside and the outside of the body.

Scatter some salt along the bottom of a wooden barrel, plastic bucket, or other watertight container that will not be affected by salt. Glass and enameled steel are good choices for containers. Do not use cast iron and stainless steel. Place the salted fish skin side down in the container with the head end toward the outside. When one layer is completed, sprinkle some salt on top and add another layer of fish. Continue layering the fish, sprinkling salt over each layer until you reach the top of the barrel. Try to make each layer as level as possible, and turn each layer at right angles to the one before it to produce a crisscross effect. Add one more layer above the top of the rim of the barrel, placing it skin side up. After a day the fish will have settled and you can put a lid on the container.

Put an extra layer of salt on the top fish to prevent the brine from becoming too diluted. Since the fish layer is above the top of the container, the pickle (salt solution) won't mix with the solution deeper in the barrel. As the salt draws fresh water from the tissue, the saltwater concentration becomes weaker. Since the salt water is denser than fresh water it settles to the bottom of the barrel. The extra salt at the top serves to keep the concentration up. Use about three pounds of salt for each ten pounds of fish.

The brine that forms is usually about 70° on the hydrometer and is sufficient to cure the fish for four to ten weeks, varying with the temperature at which the salted fish are held.

Keep the temperature of the fish low. Warm temperatures (above 65°) increase the rate of the chemical reaction called *proteolysis,* in which proteins in the fish are hydrolyzed into a soft

Fig. 32. Alternate the direction of the fish when making the layers. This produces a flatter pack and conserves space. By turning each succeeding layer 90 degrees to the one just preceeding it, the fish do not become as misshapen as they otherwise would.

mush. The softened flesh splits and bacteria enter the cracks. When the fish are piled in the barrel, the cracks close because of the weight of the fish above, sealing the bacteria from the salt. Since the salt takes approximately two days to penetrate all the way through the tissue, a sizable bacterial population can develop in the meantime.

Cool temperatures (particularly during the first few days of the salting procedure) counteract two processes: the chemical reaction of proteolysis and the textural changes it produces, and the biological process of bacterial growth and its hygenic changes.

High storage temperatures and an overly high concentration of salt cause the proteins in salted fish to become denatured: That is, the amino acids are rearranged into a less soluble form. Salt concentration above ten percent will cause proteins to coagulate and denature. One indication that the fish has begun to break down is when the skin peels away from the meat too easily.

Repacking the Fish

If the fish are to be stored for longer than ten weeks, they must be repacked. Remove them from the container and rinse them in cold water (or cold brine) to remove all the salt particles. Put the fish in the original container, trying to arrange them in the original pattern. Try not to break the fish during handling.

After the fish have been repacked, make a saturated brine (100° on the hydrometer) and pour it over them before sealing the container. The fish must *always* be covered with brine (this rule also applies to the fish before they are repacked) to avoid "rust" spots, where the fish turn yellow and pick up an off taste. Salted fish should never be exposed to air for long periods unless they are to be dried. Salted fish that are not dried or covered with brine are susceptible to spoilage from the halophiles and dun bacteria.

Check the containers periodically to be sure that the brine is covering all the fish. If kept cool, salted fish will last for months. One result of warm temperatures is that the oil drains from the fish and with it the flavor and oil-soluble nutrients. Check the salinity from time to time, since "wet" fish will lower the brine concentration considerably after packing.

In repacking, an alternative to covering the fish with brine is to pack them as before, using only one pound of salt to ten pounds of fish. The salt should draw enough moisture from the tissue to make a brine, but if it doesn't, the fish must be covered with a saturated solution.

Usually fish will shrink by thirty percent in the time period before repacking. Because fat retards salt uptake, fat, oily fish will shrink less than "thin," dry fish.

THE BRINE-SALTING METHOD

Brine salting, the second method of salting fish, is often more convenient than pickle salting. Make certain that the fish are cleaned and washed. To ensure a bacteria-free pack, dip each fish in a hypochlorite solution consisting of one tablespoon of chlorine bleach to each four gallons of water. If the water is already chlorinated, the addition of bleach is unnecessary. Set the fish aside to drain on a rack before placing them in a barrel or other nonreactive container. Pour a saturated brine over the fish and stir them gently to ensure that all the surfaces are covered and that no air pockets exist.

After a certain amount of time, the brine will reach an equilibrium level with the fluids in the fish tissue. The amount of time needed for this to occur will vary, depending on the water temperature, the thickness and oiliness of the fish and the amount of agitation. Check the brine concentration every few days. The brine and water in the fish tissue will establish an equilibrium. This causes the salt concentration to drop. When it stops dropping, pour off the brine and replace it with a new 100-percent brine. Check the brine periodically and replace it if the level drops below seventy degrees on the hydrometer. Be careful when brining large batches of fish. If the salt concentration drops too much too soon, the fish may spoil.

As in the dry-salting method, keep the fish cool until the salt has penetrated the tissue. This will hinder biological action. When they are stored later, keep them cool to slow the chemical reaction.

To soak salted fish, put it in a pan and place it in the sink. Let cold water run from the tap at a slow trickle. The faster the water runs, the faster the fish will soak out. Taste the fish periodically to determine its saltiness. If it is inconvenient to leave the pan in the sink, fill the pan with water and change it periodically.

RECIPES FOR SALTED FISH

Salted fish can be used in several ways once they are freshened in fresh water. The fish cake recipe given below is one example.

Fish Cakes

2 cups freshened salted fish free of bones and skin
6 medium potatoes
2 eggs, beaten
2 tablespoons cream
2 teaspoons onion, finely chopped, or minced garlic
A few sprigs of parsley, finely chopped
Parmesan cheese
White pepper
Melted butter and vegetable oil

Boil the potatoes in the water that the fish have soaked in. Mash them thoroughly and mix in the fish. (For a somewhat different kind of fish cake, use bread crumbs in place of the potatoes.) Adjust the amount of potatoes to the amount of fish until a fairly firm mixture is formed.

Blend the eggs into the potato-fish mixture. Stir in the cream and add the onion (or minced garlic) and parsley.

Season the cakes with Parmesan cheese. Add white pepper to taste.

Shape the fish into small cakes and fry them in a fifty-fifty mixture of melted butter and vegetable oil over medium-high heat for two to three minutes, or until the cakes are golden brown. Turn the cakes over and fry the other side.

Creamed Fish

The creamy texture and the spices in this recipe help mask the cured quality of the fish.

2 cups of freshened salted fish
1 medium onion, chopped
2 tablespoons butter
1½ to 2 tablespoons flour
1 cup milk
¼ teaspoon ground cloves

Gently simmer the fish in water for ten to fifteen minutes. Drain the water and flake the fish, discarding any bones or skin.

Sauté the onions in butter until they are transparent.

Make the cream sauce by melting the butter over low heat for a few minutes. Blend in the flour to make a smooth paste. Slowly add the milk and stir the sauce constantly for several more minutes until it thickens. Add the ground cloves and continue stirring to avoid lumps.

Add the onion and the fish, and heat the sauce until it is warm throughout. Serve the creamed fish over boiled potatoes or baking-powder biscuits.

Gravlax

This is a Swedish recipe for marinated fresh or salted salmon.

3 pounds fresh, or freshened, salted salmon, filleted and boned
Fresh dill
2 teaspoons vegetable oil
4 tablespoons fine-ground salt
4 tablespoons sugar
2 teaspoons crushed black pepper

Place the salmon fillets, skin side down, over a thick bed of fresh dill in a shallow dish. Brush the meat side of the fillets with the oil and gently rub in the salt and sugar. To guarantee an even distribution of sugar and salt, mix the two together before applying them. Sprinkle the pepper over the meat.

If there is more than one fillet, place one on top of the other, flesh touching flesh, and cover the pieces with a generous layer of fresh dill. If there is only one fillet, cover it with dill and place a piece of plastic wrap over the top to prevent it from drying out.

Let the fish marinate in the refrigerator for two to three days. If you have two pieces of fish, turn them over from time to time so that they marinate evenly.

Scrape off the dill and seasonings and slice the fish diagonally to form long, thin slices. Garnish the fish with fresh dill and lemon wedges. Serve with rye bread and a cold mustard sauce.

8

Additional Fish Products

One of the problems in working with fish is the waste that occurs throughout the different steps in processing. For example, a salmon loses twenty percent of its body weight when the head is cut off and the entrails removed. It loses an additional twenty-five percent when the sides are filleted and boned. Some of this loss is unavoidable, and often there is no use for waste, except to feed your roses and other garden members. But with imagination, you can find uses for many parts of the fish that are usually discarded.

MAKING CAVIAR

The eggs of many types of fish can be used to make caviar. Alewife, cod, haddock, salmon, shad, and whitefish eggs can be used, but the eggs of the common sturgeon, lake sturgeon, short-nosed sturgeon, green sturgeon, and white sturgeon make the best caviars. The best eggs are immature; that is, they are still somewhat small and hard. Eggs from fish that are near spawning tend to break down during processing.

Screens mounted on 2 x 2s or 2 x 4s can be used to separate the eggs from the membranes, blood, and slime. The first screen is of ¼-inch mesh. (See Fig. 33.) Rub pieces of the egg sac gently across the surface of the screen to break the sac, trapping the membrane and allowing individual eggs to fall through. Use a second screen, with a smaller mesh, to collect the eggs, allowing the blood and slime to fall through the mesh. Do not rinse the

Fig. 33. These salmon egg sacs are ready to be rubbed across the screen. Each sac contains dozens of eggs.

eggs in fresh water; it interferes with the salt cure in the next step. If the eggs must be rinsed, and they usually don't, use 100% brine. Tip the screen so that the eggs slide down the screen into a bucket, leaving the refuse behind.

Drain the eggs and mix them with a fine-grade salt. The salt used in butter works very well because of its purity and texture. The maturity of the roe and its temperature determine how much salt is needed, but an average figure is three-quarters of a pound of salt for ten pounds of eggs. Be careful not to oversalt the eggs. Use your hands to mix the eggs with the salt. If you use a metal or wooden utensil, the fragile eggs may break. Five or ten minutes of mixing should be sufficient to form a slimy foam on the top of the egg mass.

Let the eggs sit for ten minutes while the salt draws moisture from them and causes them to harden. A great deal of brine should form during this rest period. After the eggs have rested, gently mix them again with your hands for three or four minutes.

Pour the brined eggs onto a 1/32-inch mesh screen and allow them to sit until all fluids have drained off. To see if they have drained enough, press the underside of the screen. Cracks forming on top of the egg layer indicate that the roe has cured.

When the eggs are cured, pack them in sterilized half-pint jars and put them in the refrigerator. The ideal storage temperature for caviar is 26°. The eggs won't freeze because of their salt content, but they will remain fresh for several months at temperatures below 40°.

Caviar from Salmon

The large eggs from chinook (king) salmon do not make an especially good caviar; the eggs from chum (dog) and coho (silver) do. Rub the egg sac over a ½-inch mesh screen to separate the individual eggs. Rub them again over the 1/32-inch mesh screen to remove bits of membrane from the eggs. Any membranes that don't come off can be rinsed with strong brine.

A quicker method of dividing the eggs, although it requires

Fig. 34. Cured and dried fish ready to be packed into sterilized jars.

more skill, is to split the egg sacs and immerse them in hot water. Some of the eggs will be released immediately, while the remainder must be gently rubbed over the screen. If the eggs are left in the water for too long or at too high a temperature, their texture will suffer.

Make a strong brine by mixing 43 ounces of salt to each gallon of water. The brine should read 95° on the hydrometer. The eggs should soak for about thirty minutes, but the condition of the eggs and their temperature can vary the soaking time from twenty to forty minutes. Ensure equal salt absorption by gently stirring the eggs with a wooden or plastic spoon. The roe will be sufficiently cured when they begin to thicken, but before they begin to shrink and wrinkle. (See Fig. 34.)

Let the eggs drain for twelve hours and then pack them into sterilized half-pint jars. Fill the jars to the tops to keep out air, but be careful not to crush them. If kept in a refrigerator at less than

40°, they will keep for several months. Eggs should not be frozen because the egg membrane will break and a soupy liquid will form during thawing.

PREPARING EGGS FOR FISH BAIT

Salmon eggs are also widely used as fish bait. When used as bait, the eggs either can be cured singly as in caviar or, for baiting steelhead, in small clusters.

Prepare single eggs as for salmon caviar up to the point of soaking them. The brine solution for bait eggs (approximately 90° on the hydrometer) is made by mixing forty ounces of salt with from four to fifteen ounces of sugar in a gallon of water. The length of cure varies with the size of the eggs and their maturity; it should be long enough for the eggs to firm up (to prevent them from coming off the hook during casting), but not long enough for them to shrink or shrivel.

Fig. 35. Cut each skein of eggs to jar length before packing them in sterilized half-pint jars.

A dye obtained from chemical supply houses can be put into the solution to produce eggs of different hues of orange-red. Color preferences vary according to the locality, and experimentation with different amounts of dye and different lengths of dyeing time are necessary to determine the one most suitable. Use small batches for testing before committing all the eggs to a single process. Keep a record of the recipe used.

After brining the eggs, drain and pack them into clean half-pint jars. A few drops of anise oil (obtainable at drugstores) can be added to each jar to attract fish by sense of smell. Store the bait in a cool, dry place.

Fig. 36. Packing the eggs in sterilized half-pint jars.

Cluster Eggs

Cluster eggs, which are tied to fish hooks with yarn, are an easy-to-prepare bait for steelhead. Because of their firmness, immature eggs make the best bait.

Wash the egg sacs in a strong brine to remove all traces of blood and loose membrane. Then cure them in the brine given for single-bait eggs. Since the eggs are cured in a group instead of individually, they will take longer to cure, but the same conditions apply to cluster eggs as single eggs. You may want to cure the eggs just a little longer so that they will be firm enough to tie to the hook.

After curing, drain the eggs overnight and pack them in clean half-pint jars. Store them in a cool, dry place for eight to ten months. Label the jars with the date of pack and the process used.

Dry-Curing

In addition to the brining method of preserving bait eggs, dry-curing in powdered borax is a common technique. As with the brine method, the eggs can be cured in clusters or singly either in a soft or hard cure.

To soft-cure, thoroughly mix the individual eggs with plain borax in a deep bowl or on newspaper until each egg is completely covered. Pack the eggs in airtight containers and store them in the refrigerator. To guarantee longer storage or to produce firmer eggs, dry them in the open air for a short time before packing them.

Cluster eggs can be soft-cured in borax, but they are usually hard-cured. The hard cure turns the eggs a deeper red and makes them firmer. Thus they will keep for longer periods without refrigeration. These baits are easily made by covering the skeins of eggs with a fifty-fifty mixture of granulated white sugar and powdered borax. Put the eggs in an uncovered bowl in the refrigerator for one and a half to two days. The borax will absorb any fishy smell from the eggs, so it isn't a problem to store the eggs uncovered.

Remove the bait and pour off the juice that has accumulated. Rinse the eggs briefly in cold water and drain them. Cut the skeins into bait-sized clusters and let them dry and harden in the refrigerator for two days. Pack the clusters into sterilized pint jars, refrigerate or store in a cool, dark place, and don't open them until they are to be used.

PREPARING OTHER FISH ROES

Cod ovaries are simple to smoke. Thoroughly wash the ovaries in cold water, being careful to remove all traces of blood and membranes. Dredge each ovary through fine-ground salt and place them in a small tub to cure for twelve to twenty hours. Take the ovaries from the salt and give them a quick rinse in fresh water to remove the salt particles. Place the ovaries on a greased rack and smoke them for twelve hours in a 70° smoke.

Many fish roes, including herring, shad, yellow perch, trout, walleyed pike, flounder, tuna, mullet, and halibut, are edible, but roes from great barracuda, puffers (sea squab or blowfish), gar, and trunkfish are toxic. The edible roes are easy to prepare by frying. To fry shad roe, for example, wash the roe in cold water and pat it dry. Pull away the large outer membranes and prick small holes in the inner membrane. The holes allow steam to escape through the membrane as the roe cooks. Make a coating for the roe by combining flour and cornmeal. Add a small amount of salt and pepper. Melt some butter in a frying pan and add an approximately equal amount of vegetable oil to it. Bacon fat can be substituted for the oil. Fry each side of the eggs until they are brown. Baste them continuously while frying.

SALMON JERKY

One of the best products to come from a salmon (or other fish) is the meat that is scraped off of the backbone. After the fish is filleted, the meat that remains on the bone can be scraped off with a metal spoon or a round-end knife and used for fish jerky, fish cakes, or for breading and frying. The meat is also good for making fish loaf and pâté.

To make salmon jerky, collect the pieces of backbone scrapings in a bowl and pour off the blood. Mix one-third to one-half cup of saturated brine with every five pounds of meat. The brine provides the flesh with a salty flavor and slows bacterial growth. Spread a piece of hardware cloth or window screen over a metal rack (or fine chicken wire on a wooden frame) and brush it with vegetable oil to keep the meat from sticking. Drop the pieces of salted meat, a tablespoon at a time, onto the screen. Press the

Fig. 37. These backbone scrapings have been brined and put on screens to be smoked. Each dab of meat is from 1" to 1½" in diameter.

drops lightly to flatten them. Too thick a portion will not dry uniformly, while too thin a portion will dry too quickly and become crumbly.

Place the screen in the smokehouse and smoke the fish for forty-eight hours at 70°, or until it is dry and chewy. After it is smoked, the jerky should crack when it is bent, but it should not be brittle. The jerky will keep for months stored in a cool, dry place. The deep red color and hearty smoke taste make the smoked jerky a delicious treat.

FISH RECIPES

Fish Cakes

Fish cakes may conjure up images of bland, mushy, leftover fish, but these snacks can make tasty treats. Any piece of raw fish will make a good cake, but the meat scraped from the backbone is particularly good. Many Scandinavian recipes specify using ling cod in fish cakes, but some of the best are made from salmon.

For every cup of raw, chopped fish, put one cup of milk in a blender and add two tablespoons of potato flour, one teaspoon of salt, and one-quarter teaspoon of nutmeg. Add one egg and the fish and blend the ingredients until smooth. Form the mixture into silver dollar-size cakes, about three-eighths inch thick, or into one-inch balls and fry them in butter. The fish bones should be cooked in water to make a fish stock and the fish cakes then cooked in the broth. Use approximately one tablespoon of broth for each fish cake.

Another use for backbone meat is simple but surprisingly good. Add the drained pieces of meat to a bowl of flour and salt and mix them until they are well coated. Fry them in butter or olive oil and serve the flakes as a main dish.

Fish Loaf

The meat scrapings from the backbones are the most important ingredients in this flavorful fish loaf.

2 cups of cooked, flaked fish
2 tablespoons butter
¼ cup onions, chopped
¼ cup celery, chopped
2 cups whole tomatoes, chopped
1 egg, beaten
1 cup bread cubes
Salt and pepper to taste
½ cup Cheddar cheese, grated
Paprika

Remove the flesh from a salmon vertebrae. Set the meat aside so that the blood can drain while the other ingredients are assembled. Steam the backbone scrapings over boiling water. Melt the butter in a frying pan and sauté the chopped onions until they are clear. Remove the onions from the heat and add the celery and tomatoes. Stir the vegetables until they are well mixed.

Place the fish, beaten egg, and bread cubes in a large bowl and add the vegetables. Add salt and pepper to taste.

Put the mixture in a greased 9- by 5-inch casserole and sprinkle grated Cheddar cheese on top.

Bake at 350° for thirty-five minutes. Dust the top with paprika before serving.

A Fish Quiche

The Crust:

Sift and measure 1 cup flour and resift it with ½ teaspoon salt. Cut in ⅓ cup shortening with a pastry blender and sprinkle in a few tablespoons of water. Do not overwork the dough.

Lay a sheet of aluminum foil on the work surface and place the dough on it. Put another sheet of foil on top of the dough and roll it out to pan size.

Put the dough in the pan and prick it with a fork to prevent it from heaving during baking. Bake the crust in a preheated 450° oven for ten to twelve minutes.

The Quiche Filling:

1 cup steamed backbone scrapings
½ cup grated Jack cheese
4 eggs, beaten
¼ cup cream
Salt and pepper to taste

Put the ingredients in a blender and blend them briefly to form a semismooth liquid. At this point, any of several combinations of chopped vegetables can be added to the mixture to flavor the quiche and give it color. Try slices of green onion and pieces of pimento, with ½ teaspoon mustard powder. Or add sliced black olives and 1 teaspoon dill weed.

Pour the quiche filling into the baked pie shell and put it in a preheated oven at 425° and bake for thirty minutes.

Once the filling has browned, the crust may begin to overcook. To prevent this, place a sheet of aluminum foil loosely over the quiche.

Smoked Fish Pâté

In fish pâté because the type of fish used and the length of time it has been smoked (thus its degree of dryness) varies, experiment to find the proper proportions of each ingredient. Tomatoes, minced onions, and spices (dill, salt, pepper, thyme) should be ground with the trimmed pieces of smoked fish to make a smooth spread. To bind the ingredients, add milk or light cream, or perhaps a dry white wine. Worcestershire sauce is also good in pâté.

Bake the pâté in a 350° oven for thirty minutes, or until it is firm and a knife inserted in the center comes out clean.

Fish Heads

The heads of fish are a useful byproduct of the butchering process. Cut the small bits of meat from the cheeks, bread them with flour or cracker crumbs, and fry the dabs in butter. Fish markets often sell halibut cheeks (at exhorbitant prices) to be prepared in this way. The heads themselves can be cut into quarters and boiled to make fish stock, or after the quarters are cooked, the meat can be picked off the bone and cartilage and used for chowder.

Salmon Hearts

Salmon hearts, just like chicken hearts, make tasty hors d'oeuvres. Slice the hearts to form small discs about the size of a quarter. Bread each one in flour and fry them in butter.

Other Leftovers

Ragged pieces of fish that are trimmed away from fillets, bones with large amounts of meat on them, and other assorted pieces of fish can be treated like jerky (brined and laid on screens) and smoked at 70° for three to five days, or until they are dry and crunchy. The pieces with bones should then be baked until they are soft in a 350° oven with a pan of water in it, or steamed in a pressure cooker at one atmosphere of pressure.

After they are cooked, dry the bones in the smokehouse again. The smoked pieces can then be ground in a blender or meat grinder to a reasonably fine powder. The powder can be mixed with sour cream or cream cheese and spices to make a chip dip, sprinkled over a sour-cream-topped baked potato, or to cover cheese balls.

Sauces for Fish

There are many sauces that complement fish dishes. Fish loaf, fried backbone scrapings, and lutefisk are good with a well-made sauce.

White Sauce

White sauce is perhaps the simplest sauce to make, and it serves as a base for a variety of other sauces. To make white sauce, melt 2 tablespoons butter over low heat or in a double boiler. Stir 1½ to 2 tablespoons flour into the butter and blend the mixture for about four minutes. Slowly pour in 1 cup milk, or a combination of milk and strained fish stock, as you stir. Continue stirring until the sauce is thick and smooth. Add salt to taste.

The white sauce can be flavored with any of the following: cloves, bay leaves, chopped onions, parsley, chives, Worcestershire sauce, nutmeg, white pepper, or cayenne.

Mornay Sauce

It is easy to concoct a Mornay sauce by taking 1 cup of heated white sauce and adding 1 tablespoon of finely chopped onions to it. Remove a few tablespoons of the sauce from the pan and beat an egg yolk into the smaller quantity. Blend the egg-yolk mixture into the white sauce and stir in 2 tablespoons each of grated Gruyère and Parmesan cheese. Continue to stir the sauce until the cheese melts and the sauce thickens. Season with salt and cayenne.

Sour Cream Dill Sauce

Create a sour cream-dill sauce for fish by combining equal parts of fresh sour cream and white sauce. Add to it 1 teaspoon of dill weed.

Mustard Sauce

Make a cold mustard sauce by blending 3 tablespoons prepared mustard, 1 tablespoon each of sugar and white vinegar, and a little salt and pepper in a bowl. Add 5 tablespoons of vegetable oil to the sauce, drop by drop, stirring with a wooden spoon. Just before serving the sauce, add fresh dill, finely chopped onions, or chopped, hard-cooked egg yolks.

The fish products listed in this book are just some of the possibilities. Be creative with fish and don't let preconceived ideas about the uses of fish limit your approach. There is a wide variety of fish in the world and an almost limitless number of ways to use them. Properly stored and skillfully prepared, the most humble fish can be worthy of a gourmet's attention. The key to using fish products is to learn the basic techniques and then to experiment with confidence.

Appendix A

METRIC CONVERSIONS

The English System of measurement is used throughout this book.

Degrees Fahrenheit	Degrees Centigrade
0	-18
32	0
35	2
39	4
40	4
41	5
45	7
65	18
70	21

Degrees Fahrenheit	Degrees Centigrade
80	27
90	32
105	41
110	43
120	49
125	52
130	54
140	60
145	63
175	79
185	85
212	100
240	116
250	121

As the chart indicates, 0° Centigrade is equal to 32° Fahrenheit. The boiling points are 100° and 212°, respectively. To convert any Fahrenheit temperature (F) to Centigrade (C), use the formula $C = 5/9\ (F-32)$. For example, 212°F is equal to 100°C because $C = 5/9\ (212-32)$.

To measure the volume of water:

1 teaspoon = 5 milliliters
1 tablespoon = 15 milliliters
1 liter = 202 teaspoons
1 liter = 67 tablespoons
1 gallon = 3.8 liters

½ cup of salt weighs about 5 ounces, or 142 grams (1 ounce, then, weighs 28 grams, and 1 pound is 454 grams)

Appendix B

BRINE CONCENTRATIONS

The brine chart specifies the approximate amount of salt needed to make a particular brine. Since the scale on a hydrometer and salinometer is divided into 100 parts, or degrees, the readings can also be interpreted as percentages. For example, a 40° reading on the hydrometer is a brine that is forty percent saturated with salt.

Make the forty-percent brine in one of two ways: Either mix 1½ cups + 1 tablespoon of granulated salt into a gallon of water, or take 4 parts of a 100-percent solution (one that has had 4½ cups + 1½ tablespoons of salt dissolved in a gallon of water) and add it to 6 parts of fresh water.

Percent Brine (hydrometer degrees)	*Granulated Salt per Gallon of Fresh Water (in cups + tablespoons)*	*Salt per Gallon of Fresh Water (in ounces)*	*Salt per Liter of Fresh Water (in grams)*
5	2½T	1.7	13
10	¼c + 2½T	3.5	26
15	½c + ½T	5.3	39
20	½c + 3½T	7.1	53
25	¾c + 2½T	9.1	68
30	1c + 1½T	11.0	83
35	1¼c + 1T	13.1	98
40	1½c + 1T	15.5	116
45	1¾c	17.4	130
50	1¾c + 3T	19.5	146
55	2c + 3T	21.8	164
60	2¼c + 3T	24.2	181
65	2½c + 2½T	26.6	199
70	2¾c + 2½T	29.1	218
75	3c + 3T	32.0	240
80	3¼c + 3T	34.0	254
85	3¾c + ½T	37.8	283
90	4c	40.0	300
95	4¼c + 1T	43.1	323
100	4½c + 1½T	46.0	345

Glossary

Acid: A compound able to release a hydrogen ion in a solution. The acids in this book reduce the pH number to a level that indicates an acid concentration of enough strength to aid in preserving fish. The lower the pH number, the greater the concentration of acid in a solution.

Adductor muscle: The muscle of a clam that connects the two shells and holds them closed.

Aerobic: A condition in which oxygen is present. Aerobic organisms can live only when they have oxygen.

Air-dry: To dry in the open air without mechanical assistance. When a product is air-dried, it will not release any more moisture when exposed to air.

Aliphatic acid: Acids derived from fat, but also including the acids from the paraffin group of compounds.

Alkaline: A solution where the pH number is greater than 7. The higher the pH number, up to a maximum of 14, the stronger the alkaline base. Alkaline substances are often called bases.

Amino acid: Any organic acid that makes protein.

Anaerobic: The opposite of aerobic. Anaerobic organisms cannot live in the presence of oxygen.

Ascorbic acid: Vitamin C; a water-soluble vitamin used as an antioxidant.

Autolysis: Self-digestion of tissue.

Bilirubin: A reddish-yellow pigment in blood.

Blanch: To scald or parboil in water or steam.

Blower: A device with a rotating drum that produces a current of air.

Book cut: A method of cutting round-shaped fish that opens the fish like a book; the "binding" is the position of the backbone before it was removed.

Brine: A solution of salt and water.

Brine salting: A method of salt-preserving fish. The fish are packed in a container and covered with a saturated brine.

Brisling: A small herring that is processed like a sardine; also sprat, britt, or bristling.

Butcher: To cut the head off and to remove the entrails from a fish.

Butterfly cut: Similar to a book cut but with the backbone removed. When placed on a table, the still-connected halves of the fish lie flat.

Buttons: The little round pieces of bone that remain with the fish after it has been split. The buttons are formed when the ribs are cut off at the spine.

Capillary action: The action where a liquid draws itself up inside a narrow tube because of the attraction of the liquid to the solid walls of the tube.

Chemical change: A change in the chemical makeup of a substance. Converting proteins to simpler units through autolysis is a chemical change. Burning also is a chemical change.

Clostridium botulinium: An anaerobic, spore-forming bacteria that produces deadly toxins in improperly preserved foods.

Cold-smoking: The process of smoking fish at cool temperatures (around 70°) for a relatively long period.

Collar: A hard, semicircular area just behind the head of the fish that protects the gills.

Condensation: The process where water vapor changes to liquid. Condensation usually occurs when warm, moist air touches a colder surface.

Coniferous woods: Evergreen-type trees and shrubs. The leaves are needle-shaped and the wood is resinous.

Curd: The white chunky pieces that form on canned or smoked fish. It occurs where proteins dissolved in water coagulate and thicken.

Damper: A valve in a section of stovepipe for regulating the flow of air through the pipe.

Dead spots: Those areas in a smokehouse where air and smoke do not circulate. The smoke stagnates in these spots, usually in the corners.

Deciduous wood: From hardwood trees, such as alder, walnut, oak, and hickory. This is the type of wood used for smoking.

Declining rate: The second phase of the drying process. It is sometimes referred to as the falling-rate phase. During this stage, the rate of drying slows because the water can evaporate only as fast as it can reach the surface of the fish. The farther the water has to travel through the fish, the slower the rate of drying.

Dehydration: The process of removing water from food or tissue.

Denature: To modify protein by the use of heat, acid, base, or radiation to change the original properties of the protein. Toxins are denatured to make them harmless. Fish protein is denatured by improper storage.

Desiccate: To dry food by removing water.

Destructive distillation: The process where wood is broken down by heating it in a closed container without combusting it.

Dew point: The temperature at which the water vapor in the air begins to condense. If a given volume of air is saturated (holding all the air it can) at 80°, the water will condense to a liquid if the temperature drops to 79°.

Dinoflagellate plankton: Plantlike marine organisms that are present in the food chain. Some species cause "red tide" conditions.

Dorsal fin: The large fin in the center of the back of a fish.

Dressing fish: To remove the fins, tail, and thin belly strips from a fish.

Drip: The fluids that seep from frozen fish as they thaw.

Drying: Dehydrating the fish. The process of lowering the moisture content of the fish tissue to a level that won't support bacterial activity.

Dry-salting: The method of salting where dry salt is scattered between layers of fish in a container. The salt draws moisture from the tissue to form the preserving brine. Also called *pickle salting.*

Dun: A condition where bacteria grow on salted fish, creating small dark spots on the flesh.

Elasmobranch: The class of fish that includes sharks and rays.

Expanded metal: Sheet metal that is cut and stretched into different patterns. The holes may be shaped like diamonds or fleurs-de-lis.

Fillets: The piece of fish made by cutting the flesh away from the backbone. Flatfish have four fillets and round fish have two fillets. The fillets may be trimmed and skinned so that all that remains is meat.

Filter feeder: Those marine animals that gather their food by pumping seawater through their digestive system to filter out nutrients. Clams are an example.

Fine-grain salt: Salt ground into small particles.

Firebox: The chamber that holds the fire. It is often lined with firebrick. In hot-smoking, the firebox is in the smokehouse. In cold-smoking, it is away from the smokehouse.

Food-grade salt: Salt that can be used in processing food. Some salts, like rock salt, cannot be used because they contain dirt and other contaminants.

Freezer burn: The yellow color on food that indicates dehydration of the tissue. In addition to poor color, the food has a bad taste.

Freezing: The process of preserving fish by storing them at temperatures below 32° Fahrenheit. The fish solidify and create an environment that won't support bacterial growth.

Freshened fish: Fish that have been soaked in continuously changing water either to remove salt from the flesh or, in the case of dried fish, to introduce water to the tissues.

Gib: To disembowel or eviscerate fish without cutting it open.

Halophiles: Salt-loving organisms.

Head space: In canning, the space between the top of the fish and the lid of the jar.

Hook sticks: 1″ x 2″ sticks with double hooks attached along the length of the 1″ sides. The sticks are used to hang fish.

Hot smoke: The type of smoking where the fish are partially cooked while they smoke. Hot-smoked fish are more perishable than cold-smoked fish.

Hundred-percent brine: A brine that has as much salt dissolved in it as possible. A solution saturated with approximately three pounds of salt per gallon.

Hydrolyzed: A chemical reaction where molecular bonds are split and water molecules are added.

Hydrometer: An instrument that indicates the strength of a water solution by determining the specific gravity of the solution.

Hypochlorite: The salt of hypochlorous acid. A weak but strongly oxidizing chlorine-based acid.

Icing: Placing ice in the body cavity of fish and spreading ice over the outside of the fish.

Kippers: Herring cured in a specific manner with a specific smoke. The term is often used to designate any smoked fish.

Kipper-type product: Usually a piece of fish that has been hot-smoked. It is flaky and moist and has been partially cooked. The term is not limited to herring.

Liquor: The juices from clams and oysters that is contained within the shells; also *nectar.*

Lox-type product: Mildly smoked salmon. In general, the term means any fish that is mildly cured in salt and smoked for a relatively short time in cold smoke.

Marinating: Another word for pickling. Marinating is the process of soaking fish in a vinegar solution that is flavored with spices and herbs.

Medium-grain salt: Ground salt that avoids the undesirable characteristics of fine ground and rock salt. It is of a medium consistency.

Microbe: A microorganism; germ.

Mycotoxins: Poisons produced by fungus and molds.

Nail board: A board with a nail driven through it. The nail is driven at an angle to hold the head end of a fish that is being split.

Nectar: The juices held inside the shells of clams and oysters; also *liquor.*

Osmosis: The process where water passes through a semipermeable membrane (a cell membrane of fish) to equalize the water concentration on both sides of the membrane. A saltwater solution has a lower water concentration than pure water, so water would pass through the cell membrane to increase the water concentration of the salt solution. The tendency of the water to move through the membrane is called osmotic pressure.

Oxidation: To add oxygen (from the atmosphere) to a substance by a chemical reaction to form a new compound.

Oyster spat: The young of oysters that attach themselves to old or mature oyster shells.

Oyster spawn: The eggs of oysters.

Pectoral fins: The fins on the side of the fish, near the gill covers.

Pelvic fins: The ventral fin or belly fins. These fins correspond to the hind legs of an animal.

pH: A number scale that indicates the alkalinity or acidity of a solution. A value of 7 is neutral, while 0 is the strongest acid and 14 the strongest base.

Phenols: Acidic derivatives of wood tar. In diluted concentrations, they are used as disinfectants.

Physical change: Reactions or processes that produce changes in the physical state of a compound, rather than chemical changes. Changing water from ice to liquid water is a physical change.

Pickle: The brine or vinegar solution that preserves fish. The brine that forms when fish are dry-salted is often called a pickle, as is the vinegar marinade used in pickling fish.

Pickle salting: The same as dry salting fish to preserve them. The salt draws fluids from the tissue through osmosis to form a pickle.

Pickling: The process of preserving fish by storing them in an acidic solution. An acetic acid (vinegar) marinade is the most common.

Pressure canning: Heating foods above the boiling point of water. Cooking food in an airtight container with superheated steam. Pressure canning is absolutely necessary for low-acid foods.

Press piling: The method of forcing water from fish by pressing it out. The process is similar to squeezing water from a sponge.

Protein degradation: The process of breaking down protein into simpler molecules by autolysis, proteolysis, or denaturing.

Proteolysis: The hydrolysis of proteins into simpler substances. A chemical reaction that breaks down proteins.

Putrefaction: The decomposition of protein by bacteria (frequently anaerobic bacteria).

Rancidity: A chemical change where fish or other food become spoiled, giving off a foul taste or smell.

Raw spots: Places on smoked or dried fish that have not properly dried. Raw spots are usually caused by something (like oil or other fish) coming in contact with the fish and preventing air from reaching it.

Red tide: A condition where seawater turns red because of a large number of plankton. The plankton produce toxins that collect in mollusks and filter feeders, making the shellfish inedible.

Rekling: A type of dried fish produced in Norway. Flounder, halibut, cod, and other white-fleshed fish are good for drying in this way.

Relative humidity: The ratio of the actual amount of water in the air to the greatest amount the atmosphere is capable of holding at a specific temperature. The ratio is usually expressed as a percentage to indicate the extent of water saturation in the air.

Repacking: The step in salting where the fish are removed from their containers, resalted, and placed back in the containers. Repacking guarantees that the salt concentration in the brine will remain at the proper level.

Round fish: Fish as they come from the water, with the head on and the organs intact.

Round-shaped fish: Those fish that are cylindrical in appearance, like salmon.

Rust: The term used to describe salt-burned fish. The yellow-red color associated with salted fish that have not been properly stored.

Salinometer: An instrument for measuring the amount of salt in a solution.

Salmonella: An aerobic bacteria that produces the toxins causing food poisoning.

Salt-burned: Fish that have been packed in salt without a sufficient cover of brine develop a yellow or yellow-red color when exposed to the atmosphere. The colored part is said to be salt-burned.

Saltfish: Fish that are salted and then dried. Salted fish describes fish that are packed in salt without dehydrating them.

Salting: The process of packing fish in salt in an airtight container. The fish and salt combine to form a brine, or brine is added to the fish to cover it, keeping air away from the fish.

Saturation point: The point where no more of a substance can be dissolved or concentrated. The saturation point in relative humidity is 100 percent, where the air can hold no more water vapor. In terms of salt solutions, the saturation point is reached when no more salt can be dissolved in a unit of water.

Scoring: An incision or line made with a sharp instrument. Fish is scored to allow salt or other cure to penetrate the flesh.

S hooks: Hooks shaped in the form of an S. They are used for hanging fish on sticks or rods.

Shuck: The process of removing the shells from oysters or clams.

Sides: Another term for fillets of fish.

Sild: A young herring. Sometimes called *sill.*

Siphons: The tubular organs in mollusks that draw in or expel fluids.

Skein: The group of eggs that are still connected by the membrane sac. The single eggs come from the skein.

Sliming: A condition where bacteria grow on fish, especially during press piling. Sliming can also be the act of removing the slime from fish.

Smoke spreader: A board with holes in it whose purpose is to distribute the smoke evenly in a smokehouse.

Smoking: The art of preserving fish by subjecting them to smoke for an extended period of time. The preservative qualities are derived from the particles and vapors deposited on the fish and from the dehydrating effects of circulating air.

Sour: Rotten or spoiled. Fish that aren't properly prepared sour in the smokehouse.

Splitting: The process of filleting fish.

Staphylococcus: Spherically shaped bacteria.

Stockfish: Dried fish. Fish that are dried without being preserved in salt.

Strong brine: A brine that is approaching the saturation point of forty-six ounces of salt per gallon of water.

Struck (or strike): The point at which the salt has passed to the center of the fish. It is that point where all the tissue has been exposed to the salt.

Tenter sticks: Sticks used to hang fish or force them open to receive smoke during the smoking process.

Thin fish: Fish that are close to spawning, or fish that have thin muscle walls. Oily fish that have lost all their oil and juices are sometimes referred to as thin fish.

Throat: The part of the fish at the neck; often, more specifically, the ventral side of the neck.

Undercut: The cut, made on either side of the backbone, from the vent to the tail before a fish is split.

Vent: The anus of the fish, sometimes referred to as the anal vent.

Vent: The process where air is driven from the pressure canner. Venting is accomplished by letting the steam made by the boiling water in the pressure canner escape through the vent in the lid.

Ventral: The part of the fish opposite to the back, toward the belly.

Viscera: The internal organs of a fish.

Water marks: Usually black or green blotches on a salmon that has entered fresh water. On some species, the marks are red.

Wet fish: Fish that contain a high proportion of water, particularly cured fish that have been rehydrated.

White-fleshed fish: Ocean fish with white flesh, such as cod, flounder, and halibut.

Index